Wi-Fi

Digital Media and Society Series

Nancy Baym, *Personal Connections in the Digital Age* 2nd edition
Taina Bucher, *Facebook*
Mercedes Bunz and Graham Meikle, *The Internet of Things*
Jean Burgess and Joshua Green, *YouTube* 2nd edition
Mark Deuze, *Media Work*
Andrew Dubber, *Radio in the Digital Age*
Quinn DuPont, *Cryptocurrencies and Blockchains*
Charles Ess, *Digital Media Ethics* 3rd edition
Jordan Frith, *Smartphones as Locative Media*
Alexander Halavais, *Search Engine Society* 2nd edition
Martin Hand, *Ubiquitous Photography*
Robert Hassan, *The Information Society*
Tim Jordan, *Hacking*
Graeme Kirkpatrick, *Computer Games and the Social Imaginary*
Tama Leaver, Tim Highfield, and Crystal Abidin, *Instagram*
Leah A. Lievrouw, *Alternative and Activist New Media*
Rich Ling and Jonathan Donner, *Mobile Communication*
Donald Matheson and Stuart Allan, *Digital War Reporting*
Dhiraj Murthy, *Twitter* 2nd edition
Zizi A. Papacharissi, *A Private Sphere: Democracy in a Digital Age*
Julian Thomas, Rowan Wilken, and Ellie Rennie, *Wi-Fi*
Jill Walker Rettberg, *Blogging* 2nd edition
Patrik Wikström, *The Music Industry* 3rd edition

Wi-Fi

Julian Thomas,
Rowan Wilken, and
Ellie Rennie

polity

Copyright © Julian Thomas, Rowan Wilken, and Ellie Rennie 2021

The right of Julian Thomas, Rowan Wilken, and Ellie Rennie to be identified as Authors of this Work has been asserted in accordance with the UK Copyright, Designs and Patents Act 1988.

First published in 2021 by Polity Press

Polity Press
65 Bridge Street
Cambridge CB2 1UR, UK

Polity Press
101 Station Landing
Suite 300
Medford, MA 02155, USA

All rights reserved. Except for the quotation of short passages for the purpose of criticism and review, no part of this publication may be reproduced, stored in a retrieval system or transmitted, in any form or by any means, electronic, mechanical, photocopying, recording or otherwise, without the prior permission of the publisher.

ISBN-13: 978-1-5095-2989-6
ISBN-13: 978-1-5095-2990-2 (pb)

A catalogue record for this book is available from the British Library.

Typeset in 10.25 on 13pt Scala
by Fakenham Prepress Solutions, Fakenham, Norfolk NR21 8NL
Printed and bound in Great Britain by TJ Books Limited, Padstow

The publisher has used its best endeavours to ensure that the URLs for external websites referred to in this book are correct and active at the time of going to press. However, the publisher has no responsibility for the websites and can make no guarantee that a site will remain live or that the content is or will remain appropriate.

Every effort has been made to trace all copyright holders, but if any have been overlooked the publisher will be pleased to include any necessary credits in any subsequent reprint or edition.

For further information on Polity, visit our website:
politybooks.com

Contents

Figures	vi
Acknowledgements	vii
1 Why Wi-Fi Matters	1
2 Infrastructure	25
3 Home	51
4 Community	82
5 City	108
6 Problems, Prospects, Possibilities	140
Bibliography	155
Index	184

Figures

1.1 A mobile Wi-Fi hotspot, provided by the Australian internet network operator NBN during the 2019/20 bushfire season, at an evacuation centre, Bateman's Bay, New South Wales. Source: NBN Co. Ltd.   3

1.2 Internet everywhere: public Wi-Fi, Talinn, Estonia. Authors' image.   6

1.3 Square at the Centre Pompidou, Paris, France. Source: F1 online digitale Bildagentur GmbH / Alamy Stock Photo.   15

2.1 Nikola Tesla holding a gas-filled phosphor-coated wireless light bulb *circa* mid-1896. Source: Tesla Universe.   34

2.2 ALOHA terminal control unit, 1971. From Schwartz and Abramson (2009, p. 22).   37

2.3 Apple AirPort base station. © Mark Richards. Source: The Computer History Museum.   43

4.1 'Signal Code of Trampdom' in the *Kendrick Gazette* (Kendrick, Idaho), 4 June 1909.   91

4.2 Warchalking symbols. Source: Wikimedia Commons   91

4.3 'Another view of Cantenna II'. Source: Flickr/ lungstruck licensed under Creative Commons CC BY-SA 2.0 https://creativecommons.org/licenses/by-sa/2.0.   96

5.1 Immaterials: Light Painting WiFi (2011). Source: Einar Sneve Martinussen, Jørn Knutsen, and Timo Arnall, The Oslo School of Architecture and Design.   109

5.2 'A WiFi network from an 1890s apartment building spilling into the street' (Martinussen, 2011).   109

5.3 Café doors in the Old Town, Tallinn, Estonia, with a sticker promoting Wi-Fi availability. Authors' image.   124

Acknowledgements

Many people have contributed to our work on this book, and to the research projects and writings that have led to it. We would like to thank all our colleagues in RMIT's Technology, Communication and Policy Lab, and in the Digital Ethnography Research Centre. Warm thanks especially to Hannah Withers, who worked with us on this project in both its formative and concluding stages. We would also like to thank RMIT for the University's generous support for our work throughout. Thanks also to Daniel Sacchero (Easyweb Digital) and Jenny McFarland (CAYLUS) for interviews with Ellie Rennie on Wi-Fi in remote Aboriginal communities (Chapter 4).

At Polity, we wish to thank Mary Savigar and Ellen MacDonald-Kramer for their enthusiastic support of this project, their patience, and their assistance in seeing this book through to publication.

Closer to home, Julian would like to thank Jeannine Jacobson and Sam Thomas for their patience, especially when the Wi-Fi was down; Rowan would like to give a big thanks to Karen, Laz, Max, and Sunday, for their love, support, and encouragement; and Ellie would like to thank Jason Potts for giving her lots of experience with Wi-Fi troubleshooting over the years (earning her the badge for SuperTechSupport at home).

Finally, thanks to all the first pets for being excellent passwords over the years.

Julian Thomas
Rowan Wilken
Ellie Rennie

1
Why Wi-Fi Matters

When catastrophe strikes, we see communication in new ways. January 2020 was high summer in the southern hemisphere. Holiday makers, together with their smartphones, flocked to the beaches and camping grounds of the Australian coast. But, after years of drought and record high temperatures linked to global warming, the forests were on fire. Hot days, dry air, parched bush, and gusty winds created the conditions for huge, fast-moving fires. Ancient rainforests which had never experienced fire were lost. The fires isolated and burned through small townships. They left thousands of people homeless and stranded. With roads blocked by fire and fallen trees, people were trapped in small coastal communities for many days. There they managed as best they could, sheltering on beaches, making do in tents, sheds, and caravans.

Months later, when eventually the fires were controlled or extinguished by rain, the damage to people, wildlife, and the environment was immense: 186,000 square kilometres were burnt and 5,900 buildings were lost. Thirty-four people died, and an estimated 3 billion animals were killed or displaced. As well as burning bush and buildings, the fire ravaged the essential infrastructure people rely on: power lines, water supplies, roads, and communications. The fire destroyed cellular phone towers and landlines. Fixed and mobile broadband services failed, leaving homes, visitors, and businesses without internet. Destruction led to disconnection. Electronic payment systems and cash machines could not be used; health records could not be accessed.

A crisis of this kind reveals not only the central importance of modern communication networks for our safety, security, and economic and social connections, but also their fragility. It uncovers both remarkable adaptive capabilities and fundamental weaknesses. The gaps appear between our settled expectations of communication and the unpredictability of events. In the immediate aftermath of the fires, reconnection to communication was a critical priority. Stop-gap solutions depended on what was damaged and what resources could be readily used. If cell-phone towers or exchanges were undamaged but power supplies were destroyed, back-up power could be provided, sometimes with the help of the fire services. Where cellular networks were available, telecommunications companies could increase data allowances or make other concessions to users. If public phone booths were functioning, they could be reconfigured as public Wi-Fi base stations. Wireless communication networks could stand in for burnt-out fixed lines. Australia's national broadband network offered access to its satellite internet service through trucks which combined a mobile satellite connection, Wi-Fi base stations, laptops, and charging points for mobile devices.

As we finished this book, some months after the fires, a number of these mobile connection units were still in place in small towns across the fireground, parked in public spaces, reserves, and camping grounds. They continued to provide essential communications while the arduous and protracted business of recovery and rebuilding proceeded.

Events also moved on: by March 2020, and only barely after the worst of the fires in Australia, a new catastrophe engulfed us, of a kind few expected. In the space of just a year, the Covid-19 pandemic has taken over 2 million lives. In order to control the spread of the virus, governments around the world froze their economies. They closed borders, schools, universities, and businesses. Millions of people lost their jobs. Those who could work from home were required or impelled to do so. For those families that were connected,

Figure 1.1 A mobile Wi-Fi hotspot, provided by the Australian internet network operator NBN during the 2019/20 bushfire season, at an evacuation centre, Bateman's Bay, New South Wales. The truck offers free Wi-Fi and device charging. It connects to the internet through NBN's satellite service. *Source*: NBN Co. Ltd.

parents began supervising their children's online lessons, juggling school with the imperatives of work. For households without internet, the difficulties were multiplied. According to UNICEF, the United Nations International Children's Emergency Fund (2020), around a third of the world's school children were without access to remote learning, making it impossible for them to continue their education.

For us, the authors of this book, and for many others, the pandemic abruptly suspended mobile working lives, the everyday cycles of work at the office and periodic travel for meetings, research, and conferences. Working at home was always a necessary part of that cycle; now, for those fortunate enough to keep their jobs or find new ones, working space and domestic space entirely converged. New physical and functional segmentations of the home were required to make work and study spaces for everyone, from the kitchen table to the corners of rooms and corridors intended for other things. Home Wi-Fi assumed a critical role, as we rapidly came to rely on it for maintaining the simultaneous multiplicity of education, family, and social connections, as well as the everyday tasks of teaching, research, and professional communication.

The pandemic created a new 'landscape of risk' (Robinson et al., 2020; Zinn and McDonald, 2018). For those connected people able to work and shelter at home, Wi-Fi made possible a domestic bubble, a safer space offering shelter while the pandemic progressed. These people were the best placed to sustain their health and welfare during the pandemic. They could carry on without greatly exposing themselves to the risk of infection. Outside the bubble, the experiences of those without affordable communications and the skills to use them were very different. Just as the role of private Wi-Fi suddenly expanded in the home, so access to public Wi-Fi receded just as quickly. Libraries, schools, and universities closed. Cafés where students once lingered over their laptops were reduced to serving coffee to go. Many people avoided public transport if possible. Low-income families with school age children, homeless and vulnerable people, were all suddenly more socially and economically isolated by virtue of their digital disconnection. Soon after cities began to shut down, reports appeared of people working from their cars in library parking lots, attempting to use the Wi-Fi from outside.

Why Wi-Fi?

The short-range wireless networking capabilities we commonly call Wi-Fi first became part of everyday digital experience (and everyday digital folklore) two decades prior to the pandemic, when then-Apple CEO Steve Jobs famously showed off his company's new notebook computer, the iBook. The iBook of 1999 was a colourful, translucently plastic device. As a millennial design object, it broke with the blocky, grey, corporate-looking laptops of the time. The iBook invited us to see through its translucent case into the machine's internals, and its organic form and integrated handle suggested an easy portability. The built-in Wi-Fi promised a seamlessly connected future, signalling the end of the era when communication was constrained by the messy physicality of cables, plugs, and sockets.

The iBook was new, but it popularized a technology that had been in development for well over a decade, building on ideas and applications with a considerably longer history. We discuss some of these below and in the chapters that follow. From today's vantage point, Wi-Fi is no longer an emerging technology, but it is an extraordinarily successful one, now deeply embedded in everyday social and economic life. It has successively moved beyond the laptop into phones, games consoles, music players, televisions, and a suite of 'smart home' devices, from speakers to security cameras.

Like television, Wi-Fi has changed public and private spaces, from households to cafés, hospitals, and libraries. It has changed the way people work, travel, and entertain themselves, enabling the creation of new markets, new spin-off technologies, and new cultural practices. The Wi-Fi Alliance, the industry certifying body which controls the Wi-Fi trademark, estimates that, as of 2020, over 13 billion Wi-Fi devices are in use, and that 4 billion Wi-Fi devices were shipped in 2019 (Wi-Fi Alliance, 2020b). So, in the space of a few decades, the global population of Wi-Fi devices has grown to comfortably exceed the number of humans. The Alliance claims that Wi-Fi is the single most-used medium for global internet traffic, and contributes substantially to the world's 3 trillion dollar mobile internet economy (2020b). One database of publicly accessible open access networks reports that there are now over 40 million free Wi-Fi hotspots globally, with almost half a million of these descendants of the public phone booth in Indonesia alone (Wiman, 2020). Growth at the global scale has also been spectacular at a domestic level. According to one market analysis, broadband-connected households in the United States in 2018 had an average of 9.1 Wi-Fi connected devices (Parks Associates, 2018).

Wi-Fi indeed seems to be everywhere, as those big numbers suggest. But if wireless broadband once appeared magical, there is a risk that for many of us it may now seem mundane. Wi-Fi predated the take-up of 3G mobile networks in the first decade of the new millennium: it offered the first accessible form of mobile broadband. Most people's computing experiences are now both mobile and connected. We are familiar with both the convenience of Wi-Fi and its irritations: the glitchy slow-downs, the password problems, the scams, the patchy coverage, the surveillance, the highly restricted public networks, and the consequences of constant connection for work and social life. Cellular wireless services are now often faster – and sometimes much faster – than Wi-Fi. In many

Figure 1.2 Internet everywhere: public Wi-Fi, Talinn, Estonia. *Source*: authors' own.

ways, Wi-Fi has both exceeded the expectations of its early advocates and disappointed them.

Wi-Fi's fundamental capability is that it enables shared, flexible, and relatively low-cost access to the internet, a valuable resource. This gives rise to a set of distinctive attributes, and these are at the heart of both the extraordinary successes of Wi-Fi and its failures. As we describe, Wi-Fi is an unusual form of network infrastructure, which augments and sometimes substitutes for other networks, while proving resistant to the power of both internet platforms and large service providers. In households, communities, and cities, Wi-Fi can work as a gap filler and a network extender. It does not rely on cutting-edge technologies or high-end processors, and Wi-Fi chips are produced in huge numbers, so the hardware is cheap. Its transmissions use the shared, publicly available spectrum, so users do not bear the costs of exclusive commercial spectrum licences. It is usually deployed on the edges of communication networks, within households and public spaces, by both end users and internet access companies. In telecommunications-speak, Wi-Fi is a 'last mile' technology, which can be provided, managed, and adapted by internet users themselves, whether these are families, institutions, or local communities. By the same token, the deployment of Wi-Fi doesn't directly change underlying network infrastructures, such as the distribution or ownership of high-speed cables and switches. Nor does it change market structures, policies, or pricing models which substantially determine where and how people can connect. This means that Wi-Fi on its own is unlikely to bridge the digital divide or equalize the social distribution of digital resources. Despite the hopes of some its early advocates, Wi-Fi has not displaced commercial mobile networks. Nor has it created an open, internet commons.

For most users, Wi-Fi remains a cheap consumer 'add-on', with base stations often built into inexpensive internet access points provided by internet service providers. The low-cost model, a key reason for Wi-Fi's success, probably also explains relative underinvestment in Wi-Fi compared to other wireless technologies. Newer modes of wireless connection, including the higher-speed cellular broadband services marketed as 5G, now promise higher speeds and greater security, alongside more lucrative returns for telecommunications firms, and a new upgrade cycle for smartphone manufacturers.

So why a book about Wi-Fi at this time? The experiences of 2020 underline some of the reasons why Wi-Fi matters, and is likely to matter more in the future, and why we need to understand it better than we do. When we think about how Wi-Fi has been used in the two emergencies, the bushfires and the pandemic, we can readily compare the responsive problem-solving demonstrated in the aftermath of the Australian fires with the adaptability of connected people and organizations during the pandemic. In both cases, Wi-Fi appears as a means for restoring capabilities which had been swept away, a means for managing exceptionally difficult circumstances. For those sheltering from the fires in remote townships, Wi-Fi was deployed in new ways to provide emergency support and relief for everyone with a device to connect. Reconnection was critically important for both stranded urban holiday-makers and the residents of small coastal towns, regardless of the considerable social and economic differences between these populations. Wi-Fi services in the wake of the bushfires were usually offered by telecommunications companies, which used Wi-Fi to stand in for damaged infrastructure. For those sheltering from the pandemic at home, Wi-Fi was also an essential digital resource – a means for continuity in work, education, entertainment, and social links. In the case of the pandemic, Wi-Fi was usually self-provided, in order to extend networks within households. In these circumstances, we have also seen sharply differentiated social consequences, a magnification of digital inequalities. For those without vital digital connections at home, the risk of exposure to the virus also increased, just as the absence of connectivity increased the risk of the fires.

This book shows how, in ways and circumstances other than catastrophes, Wi-Fi continues to provide vital connections. At the same time, Wi-Fi changes the way people connect with each other, media, and digital services. Internet scholars have written about how the internet 'reconfigures access' to resources (Dutton, 2005). Online news services, for example, may reinforce people's interest in the news, by making news content more readily accessible; they can also change the kind of news people encounter, by presenting alternative sources of news. Wi-Fi invites us to consider how a flexible and affordable wireless medium may reconfigure access to the internet itself, both by making the internet more accessible across diverse physical and social locations, and by changing the ways in which people use it. The fact that Wi-Fi augments and extends networks from their edges should not lead us to underestimate its significance: it is possible to change the internet from its edges. Just as Wi-Fi is now enabling the proliferation of connected devices in households, a decade ago Wi-Fi played a key role in the evolution of smartphone ecosystems, providing a low-cost parallel network ideal for backups, downloads, system maintenance, synching, and all those data-intensive tasks best kept off more expensive cellular networks.

The events of 2020 give us some clues as to how this 'reconfiguring' works. Wi-Fi introduces a plasticity to network connections both within specific spaces and situations, such as households or cafés, and in wider public, institutional, and community settings. It does this in an unusual set of ways. We can think of Wi-Fi as 'entangled infrastructure', because its applications and utility are so dependent on their social and locational contexts. Wi-Fi is inexpensive to build into devices, and it provides access to the cheapest data available – usually from fixed broadband connections rather than cellular data. These qualities help us to deal with a whole range of urgent and contemporary problems, from the demands of home-based schoolwork to the communication needs of people in both extraordinary and everyday difficulties.

Wi-Fi therefore reminds us that the internet need not only be about corporate software, national rivalries, and vastly powerful platforms. It can also be successfully designed for cheap devices and open standards. However, the plasticity of Wi-Fi is not unlimited. Larger-scale network infrastructures, market dynamics, and public policy settings all play substantial parts in determining where and how people can connect. Despite the flexibility and popularity of Wi-Fi, internet access remains a scarce and expensive resource in many situations and places. While climate and health disasters underline the contingencies and fragilities of the communication systems many of us take for granted, everyday access to inexpensive, reliable internet is a daunting problem for large numbers of people, especially – but not only – in low- and middle-income countries. Mobile broadband has extended access to digital services and participation in the digital economy, but data costs remain high. According to the Alliance for Affordable Internet (2019), although progress is being made in some countries, the world is still decades away from universal, affordable internet access. Moreover, the network effects of the internet mean that, as more people are connected, the costs of disconnection – those disadvantages incurred by people who are wholly or partially excluded – also increase.

According to the International Monetary Fund (2020, p. xv), the world after Covid-19 is likely to be poorer and more unequal for many years to come. The pandemic has reversed global progress in reducing poverty, with only a protracted and gradual recovery expected. If we think about the impact of the pandemic on digital inequality, we see a particularly fluid and challenging dynamic. Governments and businesses are responding to Covid-19 by hastening the transition to online services. While digital transformation has many benefits, it also magnifies the problem of digital inequality – a problem with no simple fix, and many dimensions: it involves access to networks, devices, applications, and content; the cultivation of a diverse range of skills and capabilities. Digital inclusion is also about affordability – what proportion of people's incomes do we expect them to pay for essential communication and services? Wi-Fi networks have the potential to address directly problems of access and cost, and can contribute indirectly to boosting skills and capacities. This is why, in August 2020, the South Korean government announced plans to install 41,000 free public Wi-Fi hotspots by 2022, and to upgrade 18,000 older installations (Cho, 2020). It appears that Wi-Fi will continue to matter, and its role may grow in importance.

Wi-Fi through past and present

In the chapters that follow, we explore the historical trajectories of Wi-Fi in order to illuminate its present significance. We discuss Wi-Fi's deep foundations in twentieth-century theories of wireless communication; its more immediate origins in the 1970s and 1980s, in wireless network experimentation and spectrum policymaking; its emergence as a focus of public and commercial research and development in the 1980s and 1990s; and its subsequent status as an evolving set of technical protocols supporting an accelerating proliferation of devices and 'smart' technologies. Our approach throughout is not to focus on the technical aspects of Wi-Fi – we note that the relevant standards in any case comprise a large and evolving group of technologies – but on its social and institutional contexts, its uses and applications.

We have already begun to sketch the place of Wi-Fi in contemporary digital experience. We now turn to a closer consideration of what its history tells us about the significance of Wi-Fi in its many guises – as marketing strategy, as technical protocol, as open industry standard, as public utility, and as intellectual property. Wi-Fi raises intriguing questions: about the prominent visibility of this embedded, mainly hidden form of infrastructure; about the control and ownership of Wi-Fi's open standards; and about the place of Wi-Fi between the commercial tech industries, public utility, and the worlds of low-cost community and domestic networks. In order to address these questions, we can draw on both recent developments and some salient lessons from Wi-Fi's complex past.

Wi-Fi is a brand

When Steve Jobs unveiled the iBook laptop, he didn't talk about Wi-Fi – the wireless networking features were branded with an Apple trademark, 'AirPort', conveying the idea that these industry standard capabilities would be 'first and best' on Apple's machines. As other firms began to build those same capabilities into many other computers and base stations, the AirPort name inevitably became one of many used to market wireless networking gear. What became known as Wi-Fi was generally designated as '802.11', the number given to the relevant family of wireless standards developed for local networks within the Institute of Electrical and Electronics Engineers, known as the IEEE.

For those manufacturers and developers keen to promote the new standards, several issues were quickly apparent. Jobs emphasized the fact that AirPort used the new industry standard, and therefore would work with a whole array of devices soon to appear. But the 802.11 standards were complex and wide-ranging, with the result that not all compliant devices using the same standard were assured to work together, and the IEEE's role in specifying the agreed standard did not extend to testing devices for compliance. Further, 802.11 was, as we have noted, a family of standards, with each iteration given a specific alphabetic suffix. The standard used in Apple's 1999 iBook and other early consumer systems was 802.11b, to be followed in time by 802.11g, 802.11n, 802.11ac, and many others. These different versions of Wi-Fi all involved significant improvements, but the nomenclature was difficult to follow or comprehend for those without specialist knowledge.

A new trade organization emerged, the Wireless Ethernet Compatibility Alliance, to promote the new wireless networking and certify that devices would work together. For this purpose, a new name was required. Interbrand, a transnational marketing consultancy with previous successes including Prozac and oneworld, was commissioned. Interbrand conceived the 'Wi-Fi' name, together with a logo that borrowed (or appropriated) familiar yin-yang symbolism. The point was plainly to synthesize a brand, something that could be registered, licensed, and controlled through trademark law. The alliance itself became the 'Wi-Fi Alliance'. The name 'Wi-Fi' was coined in part because it could be readily trademarked – no-one else used it, nor could it be confused with anything else. It was an entirely arbitrary name which meant nothing. The word did play with 'Hi-Fi', an abbreviation for 'high fidelity' with a certain retro cachet from the world of consumer audio. However, the evidence is that the Wi-Fi name was not intended to signify 'wireless fidelity', or be an abbreviation for anything. The Alliance nevertheless confused the issue by adopting for a time the slogan 'the standard for wireless fidelity' – a formula that was developed after the name had been chosen, and meant very little. It was noted that no-one knew what wireless fidelity was, and the Alliance was not a standard-setting body (Doctorow, 2005).

The Wi-Fi Alliance currently controls around fifty Wi-Fi-related brands (Wi-Fi Alliance, 2020a). While the trademarked name plays a critical role in stabilizing a complex and evolving group of technologies, it is also surprisingly multivalent itself. Just as Humpty Dumpty once reserved for himself the right to decide what a word meant, the Wi-Fi Alliance can decide what Wi-Fi means in any given context. The Alliance's current vision is 'connecting everyone, everything, everywhere'. From hotspots to encryption and set-up systems, the Alliance's

Wi-Fi trademarks cover a remarkable range of applications and uses, as well as the many versions of the main networking protocols. The brands are of two main kinds: those for public use – such as the generic 'Wi-Fi' name itself – can be used by anyone to describe or refer to Wi-Fi products. These are licence-free, subject to a small number of requirements and prohibitions, including rules about how the word should be capitalized and hyphenated. Then there are the certification marks, exclusively for the use of Alliance members, intended to function as a 'seal of approval' for products guaranteeing interoperability, security, and compliance with relevant protocols. These are subject to strict rules and prohibitions. Meanwhile, the Alliance's branding strategy has continued to evolve, and the reach of the Wi-Fi brand has continued to expand. For many years it was used alongside the IEEE's 802.11 alphabet soup of different versions, so products using the Wi-Fi name and logo would also specify compatibility with '802.11ac' or other versions. In 2018, the Alliance began a retrospective rebranding, known as 'generational Wi-Fi', requiring the different iterations of Wi-Fi to be rebadged as 'Wi-Fi 4', 'Wi-Fi 5', and so on. The Alliance's documentation draws an explicit comparison with the effective marketing of 'generational' cellular technologies such as 4G and 5G.

This background underlines the intangible nature of Wi-Fi, but also points to ways in which the Wi-Fi Alliance uses the trademark system to exercise considerable control over the wireless networking ecosystem. While the IEEE's 802.11 standards are open for licensing, any use of the Wi-Fi name involves an additional layer of control through the Alliance. The Alliance justifies the branding of Wi-Fi on the grounds that it gives consumers confidence and peace of mind regarding the interoperability, safety, and security of networks and devices. It is clearly also a marketing strategy, shaped by a sharp appreciation of the competitive pressures in digital networking.

We can recall here the persistent criticisms from social scientists, cultural critics, and policymakers: brands can be used to raise prices, reduce competition, expand market power, and appropriate common property (see, for example, Coombe, 1998; and the discussion in Lobato and Thomas, 2015, ch. 6). The Wi-Fi brand, and the marketing strategies associated with it, may well be vulnerable to objections along these lines. But in this case the brand also plays an institutional role, mediating between the market and the technical agreements co-ordinated through the IEEE. Further, as a highly successful global tech brand which is not owned by a transnational corporation, the brand also signals Wi-Fi's double-sided orientations towards both commercial markets and public goods. Those interested can read more than connectedness into the yin-yang symbolism. Despite all that, the success of this brand only goes so far. Wi-Fi's successful progression from the home into the city streets has been followed by a proliferation of informal Wi-Fi signage, often far more widely used than the licensed logos. In the vernacular, we find Wi-Fi spelt in almost every combination of unauthorized ways, capitalized, lower-case, or unhyphenated.

Figure 1.3 Square at the Centre Pompidou, Paris, France. *Source*: F1 online digitale Bildagentur GmbH / Alamy Stock Photo.

Instead of the stipulated symbol, a generic wireless symbol depicting the radiation of the signal is ubiquitous. The Alliance may have created the name and the logo, but the do-it-yourself ethos of Wi-Fi now extends to its branding.

Wi-Fi combines new and old technologies

Wi-Fi is best approached not as a single technology, but as a large and diverse group of technical innovations which have been brought together under a single banner. These function as agreed protocols – rules for critical functions such as encryption, addressing, error correction, channel spacing, and power outputs. These are assembled for an agreed purpose – local wireless networks – which itself is likely to evolve, and they address a changing array of problems. Novel techniques are included, but one of the reasons for Wi-Fi's low cost is that it combines many pre-existing innovations, some of them already covered by well-known industry standards. So Wi-Fi makes full use of the tricks used by wired ethernet networks for handling data packets. It uses the internet's underlying protocols for directing data flows, together with 'spread spectrum' and 'frequency hopping' techniques for sharing radio frequencies, which are also used in other wireless communication systems, such as Bluetooth for short-range devices, and GPS for satellite navigation.

In these respects, Wi-Fi is a good example of Brian Arthur's observations about the piecemeal, combinatorial aspects of technological evolution (Arthur, 2010). The most significant Wi-Fi innovations involve the assembly of many layers of techniques and practices, the documentation and stabilization of the system through standard-setting, and the integration of this bundle of technologies into hardware and software. Standards and marketing play a major role in market formation, creating economies of scale for further innovation. Companies such as Apple are then well positioned to champion new standards.

The critical achievement for any emergent technology such as Wi-Fi is the overall assembly of new and old techniques, combined with consensus about shared objectives. The next stage is a protracted process of internal and incremental innovation, whereby particular components within the assembly are replaced with marginally improved or revised versions in response to known and anticipated problems. After all, Wi-Fi does something inherently difficult: it transmits and receives large volumes of data from multiple users using shared radio frequencies in spaces that are not designed for the purpose. The problems are manifold: wireless networks must deal with interference from devices using the same frequencies; they must provide secure communications; within buildings, transmissions need to deal with physical objects, especially walls and floors, which signals can bounce off or fail to pass through. Revised and faster versions of Wi-Fi need to deal with devices using older versions: do these slow down the entire network? As additional and different kinds of connected devices appear – such as phones, tablets, and smart TVs – how do we manage the multiplying number of connections? And, as more of those devices are battery powered, and as energy consumption across the network becomes an increasingly pressing question, how do we increase the overall efficiency of the system?

The recent history of Wi-Fi is often framed around a progressive narrative about increasing network speeds, where each new version of the technology generates an impressive leap in the claims made about maximum bit rates – claims which rarely translate directly into reality. Although improvements in speed are important, the many iterations of Wi-Fi (and the resulting alphabet soup) are better understood as an accretion of new technical solutions within the overall assemblage, together with evolutionary improvements in the core capabilities.

Wi-Fi's histories are diverse and contested

Because Wi-Fi is a retrospectively synthesized collection of technologies, it is also inevitably a teleological invention. From Nikola Tesla's ideas about universal wireless communication to Hedy Lamarr's wartime research, the Wi-Fi narrative encompasses many histories, and projects an inevitable progression towards a connected future. The origin story then reaches considerably further back in time than the lifetime of Wi-Fi itself. It follows that, when we assign agency to key actors in that origin story, we need to remember that, despite some extraordinary instances of technical imagination, Wi-Fi as we now know it was not the objective or intention. For example, as we describe in later chapters, Wi-Fi's antecedents and constituent elements were, like Tesla and Lamarr's breakthroughs, not conceived for domestic purposes: their lineages were academic, scientific, and commercial, and their wireless capabilities were responses to problems concerning communication at different spatial scales. The first computer network to use radio communications was ALOHANet, deployed at the University of Hawaii in 1971. ALOHA used UHF signals to connect a university distributed across an oceanic archipelago, linking users across the islands to a central computer. In the late 1980s, NCR researchers in the Netherlands created WaveLAN, a local wireless network designed for communication between cash registers in a retail space. That technology was then adapted in 1993 to build a campus network at Carnegie Mellon University. Radio astronomers at the Commonwealth Scientific and Industrial Research Organisation (CSIRO), an Australian publicly funded research and development body, devised new systems for differentiating radio signals in order to detect gravitational waves from black holes. CSIRO subsequently recognized the potential to apply the same techniques in local wireless networks, and developed a chip incorporating their approach.

Histories of Wi-Fi circulate widely in scholarly texts, popular science, and technology journalism. The stories of where Wi-Fi came from and why it has become so successful diverge along familiar planes of cleavage, producing different protagonists and chronologies. A few examples: in *Wired* magazine, we find Chris Anderson's celebrations of Wi-Fi's 'revolutionaries', the great disruptors of telecommunications and internet access (Anderson, 2003). The big story here is not just Wi-Fi, but the whole idea of 'open spectrum', a de-privatization of spectrum licensing to allow a comprehensive development of digital radio technologies. Here, the 'radical pioneers' are American technology activists, entrepreneurs, 'bandwidth pirates', and engineers, but the story also celebrates the role of political and government leaders. In contrast, the community networking field displaces the narrative centre from Silicon Valley to remote communities and dense urban spaces. Alex Hills's *Wi-Fi and the Bad Boys of Radio* (2011) is a case in point: the protagonists are not the bad boys (these are the unpredictable radio waves) but technologists with a public vocation. In *The Economist*, we find another perspective again, a recognition of the institutional and technical complexity of the history, with a strong emphasis on the facilitating role of government. In this context, the decision of the US Federal Communications Commission (FCC) in 1985 to allow unlicensed use of the 900 MHz, 2.4 GHz, and 5.8 GHz bands of the radiofrequency spectrum is the vital foundational initiative. The visionary work of FCC engineer Michael Marcus is especially notable.

At the formal level of institutions, history also matters, but it is documented in different ways and in different contexts. Wi-Fi's key technical components and their lines of descent are carefully registered and adjudicated through the exacting work of IEEE's standard-setting committees, a key element in the association's broader formalizing and standardizing role in the technology industries. IEEE's standard setting is designed to establish grounds of consensus among key industry players to promote interoperability and reliability, and to reduce product development timelines and costs. IEEE decision-making is itself highly structured and closely aligned to World Trade Organization regulations. It produces open standards but not open-source or free-to-use technologies. Notwithstanding the fact that Wi-Fi makes use of unlicensed parts of the spectrum, the IEEE standards include many technical features that do need to be licensed by device manufacturers. The commitment of the participating patent owners is only to license these on fair, reasonable, and non-discriminatory terms. IEEE standards committee deliberations, decisions, and supporting material are all carefully organized and recorded in considerable detail.

A different kind of retrospective accounting occurs through legal institutions. Conflicting versions of Wi-Fi's history have been litigated extensively, producing a substantial archive of testimonial and documentary evidence. Many parties now claim to have 'invented Wi-Fi' or made other decisive contributions to it. The rich and detailed collection of essays edited by Wolter Lemstra, Vic Hayes, and John Groenewegen (2011) on the 'innovation journey' of Wi-Fi concentrates on the development of WaveLAN. The book is authoritative, pluralistic, and wide-ranging, but it makes no mention of the CSIRO research referred to earlier, which has been celebrated in Australia as the 'invention' of Wi-Fi. The omission may not be surprising, given that recognition for CSIRO's contribution – some smart signal processing to address the 'multipath' problem (signals bouncing off walls) – was hard-fought and remains contentious. CSIRO received substantial royalties for the patents concerned only after more than a decade of litigation in US East Texas courts. The dispute was framed as a struggle between two entirely different versions of Wi-Fi history. The US tech website Ars Technica covered the settlement of the case in a state of disbelief:

Why is the history of such an invention in dispute? The premier world engineering institution, the IEEE, created a working group for the evolving 802.11 wireless standard in 1990, a full *three years* before CSIRO filed for its key wireless patent. The group voted repeatedly on which way to go forward and produced heaps of records, but CSIRO didn't even participate in the 802.11 committee. The group published the first 802.11 standard in 1997 and CSIRO came forward years after the fact. ...

A ubiquitous technology that exists because of standards – because of widespread cooperation, essentially – has been re-cast as a story of [a] noble group of hero-inventors, ahead of their time, overcoming the non-believers in court. (Mullin, 2012a)

Ars Technica was particularly concerned by the representation of the CSIRO work in Australia, with the organization listing the Wi-Fi research as its single most important discovery, and the team receiving a number of important prizes and awards. CSIRO's patent features in *A History of Intellectual Property in Fifty Objects* (Healy, 2019), a recent survey of significant innovations and their legal destinies. Here Terry Healy presents precisely the narrative of heroic invention which Ars Technica complains about: brilliant researchers pursuing an independent path. Their legal victory, so the story goes, is a victory for 'research', and for a small group of outsiders, remembering that the original work was done by radio astronomers, rather than electrical engineers. Disparaged by a self-interested club of US tech companies, these scientists in an obscure antipodean government science organization nonetheless made a significant contribution.

The merits of this case aside (and the matter was settled, so no judgment was made), it is notable that, in fact, very little of the Ars Technica argument is inconsistent with CSIRO's position, which was not that their scientists had invented Wi-Fi, but that their innovations had substantially improved it, making it fast enough to substitute for wired ethernet networks and therefore (the argument went) to become as successful as it did. Nor does patent law require proof of copying or piratical intent: infringement simply involves unlicensed use of the protected invention.

The Ars Technica position is that it is unconscionable that a very substantial interest in the technology could be assigned to a non-participant in the IEEE process. There is a strong pragmatic point at stake here about the strategies the IEEE deploys for the social organization of innovation. The standardization process was in essence a consensus-building approach, and there can be little doubt that the successful achievement of consensus – after many years, in the case of 802.11 – was critically important to the later success of Wi-Fi. But these processes, no matter how exacting, are never entirely inclusive, and cannot preclude contestation in the legal domain over ownership and control. While the IEEE managed the formative process, it could not control the consequences.

If the contested claims at the heart of the CSIRO patent dispute are understood as arguments about history, they are likely to be incommensurable rather than finally resolvable. The answer to the Ars Technica question – why is the history in dispute? – lies in the fact that Wi-Fi encompasses many histories, and it does so because of its distributed character. Wi-Fi is not like our smartphones; this is not a platform owned by any single entity or controlled by anyone. (Even CSIRO's patents are now expired.) It is not – so far at least – like 5G, which now figures in global economic and political rivalries. Huawei is banned from providing 5G equipment in the United States and some other Western countries, but no such prohibition applies to its activities in the Wi-Fi domain (including telecommunications). Wi-Fi has avoided a strong association with any government, despite the contributions of public bodies and public policy to its emergence. For these reasons, there is more at stake in disputes such as the CSIRO case than just patent royalties. They remind us how important wireless communication is to the objective of a more mobile and inclusive internet. They remind us that wireless communication need not be framed as national technology strategizing. They remind us that the internet can be substantially built around basic, low-cost hardware, and commonly agreed standards available to everyone.

Overview of chapters

Wi-Fi is an 'entangled infrastructure'. Our account of its social significance proceeds from the observation that its uses, and the debates these generate, are highly contextual. What Wi-Fi does and why it matters depends on where you are, in spatial, social, and institutional terms. So, in this book, we approach Wi-Fi across a series of different kinds of locations and at increasing social scales – the household, the community, the city. In all these contexts, we describe the transformative impact and animating potential of Wi-Fi as a distinctive form of infrastructure. We show how in different ways, Wi-Fi works not only as an occasionally essential digital service, but also as a powerful symbolic resource. Wi-Fi is an idea as well as a marketing phenomenon and technical construction, and much discourse about Wi-Fi retains a promise of the future, just as the memory of Jobs's 1999 showmanship persists. Chapter 2 shows how Wi-Fi's infrastructural dimensions and historical trajectories reveal a strong aspirational element. The proponents of Wi-Fi project ideas about communication possibilities: flexible, inclusive, inexpensive, grassroots technology, designed for the scale of households and communities. Wi-Fi's symbolic power often works to expand digital participation. Public Wi-Fi fills gaps where other services fail, but it can also crowd out more ambitious strategies for an inclusive internet.

Our discussion then moves to the household. Internet scholars have written extensively about the domestication of certain technologies – the means by which they are appropriated and adapted into everyday life, and especially the private world of the home (Haddon, 2006). Those studies have rarely included close consideration of Wi-Fi, but in Chapter 3 we look at the continuing mutation of domestic Wi-Fi. Wi-Fi complicates standard accounts of domestic technology: we consider the ways in which, while being appropriated into the home, Wi-Fi itself may be said to domesticate households. In Chapter 4, we focus on initiatives at the level of communities, returning to the aspirational dimensions of Wi-Fi, examining the conditions of possibility for community networks, their modes of control, and the forms of agency they offer. Chapter 5 locates Wi-Fi in the urban history of communication, an addition to the enveloping 'hertzian space' of urban radio traffic. Urban Wi-Fi is a sharply contested space, where we find influential but sharply divergent visions for communication futures. These include competing understandings of 'free' Wi-Fi, municipal Wi-Fi, and the Wi-Fi-enabled 'smart city'. We then conclude the book with reflections on Wi-Fi's current problems and future prospects.

2

Infrastructure

In 2013, Wayne Dobson, the owner of an ordinary house in Dallas, Texas, got so fed up with being accused of stealing people's mobile phones that he posted a sign saying 'No lost phones' on his front door (Coldewey, 2013). People were mistakenly being directed to his house by their finder app. Three years later, Christina Lee and Michael Saba, living in Atlanta, Georgia, found themselves in a similar predicament. On one occasion, the police arrived looking for a missing teenager whose phone had been traced to their house (Holley, 2016). In both cases, neither households possessed the missing devices and the journalists who covered these stories were unable to solve the mystery other than positing that it had to do with the location of cell towers. Eventually, IT sleuths from the podcast Reply All concluded that, in Christina and Michael's case, the problem stemmed from the fact that they lived in an area where very few homes had Wi-Fi connections (Reply All, 2016). Finder apps work by triangulating GPS data with Wi-Fi data. Because theirs was the most active wireless device in the area, finder apps were leading people to the couple's home, like a lone beacon on an empty landscape.

These stories illustrate the entangled nature of Wi-Fi infrastructure. Off-the-shelf wireless products interact with a range of other infrastructures – in this instance, the connection to household wired internet that the owners had set out to install, as well as geolocation infrastructures they had no control over. Wi-Fi is not just hardware components (receivers, transmitters, network interface cards); it is also the software that carries packets, the standards that determine how different devices and software operate together, and the rules that permit legal transmission over certain frequencies across electromagnetic waves of the atmosphere. These elements interact with backbone internet services, electricity grids, and the built environment. They also make Wi-Fi accessible, fragile, and corruptible.

In addition, Wi-Fi possesses an affective and aspirational dimension in that it is invoked in policy as a solution to internet access where other internet infrastructures are too costly. In rural and development contexts, Wi-Fi programs can mask digital exclusion by suggesting the presence of internet infrastructure, when in reality it can offer limited access. In this chapter, we begin by considering the particular dimensions of Wi-Fi as an infrastructure and look back at its development. We then look at Wi-Fi in remote settlements to explore the political dimensions of Wi-Fi in digital inclusion agendas.

What is an infrastructure?

Infrastructures are loosely defined as 'matter that enable the movement of other matter' (Larkin, 2013, p. 327). They are the foundations upon which systems function, such as the tracks on which trains run, the pipes through which water flows, and grids that direct electricity supply. The word itself originated from the military, used to describe fixed facilities in the context of what was required for particular services (Edwards, 2003). Even when they are broken, we think of infrastructures for what they should be doing rather than their standalone prefabricated 'thingness'. Infrastructures are in this way known for what they achieve, making them relational and active.

Let's take a very common infrastructure: a road. Roads influence our ability to move from one place to another, and they encourage some behaviours over others (such as a road built for driving rather than cycling or walking). The material form becomes an infrastructure when it creates the terms on which social practices occur, binding us to the physically possible (Berlant, 2016). The importance of an infrastructure will also change depending on who is using it, what they are using it for, and whether there are alternatives. Depending on the materials used, the degree to which it is maintained, or if it changes as it crosses a territory, a road can signify whether a place is reachable, remote, developed, or undeveloped. Some infrastructures evolve out of use over time, such as holloways – roads that were never built but created through use over thousands of years. Routines and enterprises are built through infrastructures, but they can also have an affective power – relating to emotions, memories, and feelings – such as a road evoking particular journeys to a person who travels on it (the road to work versus setting off on a holiday adventure). Because roads can help us to understand people's mobility and the flow of resources, they have been studied as 'catalysts to economic activity, urban redevelopment and cultural exchange' (Merriman and Jones, 2017). Studying infrastructure can reveal specific and tangible features of a society, including how things are done and where benefits accrue.

Star and Ruhleder (1996, p. 113) identified particular dimensions that can be used to define infrastructures:

- *Embeddedness*: 'Infrastructure is "sunk" into, inside of, other structures, social arrangements and technologies';
- *Transparency*: 'it does not have to be reinvented or assembled for each task, but invisibly supports those tasks';
- *Reach or scope*: 'has reach beyond a single event or on-site practice';
- *Learned as part of membership*: 'outsiders encounter infrastructure as a target object to be learned about' and acquire 'naturalised familiarity with its objects as they become members';

- *Links with conventions of practice*: 'Infrastructure both shapes and is shaped by the conventions of a community of practice';
- *Embodiment of standards*: 'infrastructure takes on transparency by plugging into other infrastructures and tools in a standardized fashion';
- *Built on an installed base*: Infrastructures are often overlaid onto other infrastructures in ways that can be convenient or limiting;
- *Becomes visible upon breakdown*: We tend to notice infrastructures more when they break down.

We return to these points in relation to Wi-Fi throughout this chapter, as a framework for thinking through what Wi-Fi is and does.

Wi-Fi is self-provided

Wi-Fi is unlicensed and often self-provided, which has consequences for how we experience it. You do not need permission to establish a Wi-Fi hotspot, whereas you would definitely need permission from various authorities if you wanted to establish your own mobile or satellite service. Most people can set it up without knowing much about how it works, and connect to other hotspots through basic instructions that become learnt behaviours.

When we wish to connect to Wi-Fi, we look for a common set of characteristics (a signal, a password, an icon on a screen or in settings). We repeat this same set of procedures every time we wish to connect between our own devices and to different networks. The *transparency* (Star and Ruhleder, 1996) of this particular infrastructure – as in something that does not have to be assembled each time it is used – is only partial. With Wi-Fi, there is an assembling process that we undertake when we perform steps to establish a hotspot or get connected, but we become familiar with these steps and repeat them for each network we encounter.

Another aspect of being self-provided is the visibility of Wi-Fi networks. Infrastructures are often discussed as technical systems that are taken for granted – the connective tissue that makes things run and which we cease to notice (Edwards, 2003). This is not the case with Wi-Fi. We consciously look for Wi-Fi, which becomes visible only when we are within range. We are reminded of its limits every time we become disconnected due to a weak signal. Many networks are named after the business or public location that administers them, made recognizable for a defined group of people who have permission to access it. Some remain as a cryptic string of letters issued by a manufacturer, while others are re-named by users to send a message. One hotspot we came across used the naming feature as an overt instruction. It read: 'Mom click here for internet'.

Studies of computing infrastructures have shown that they often require a high degree of flexibility or customizability in order to develop, yet also require fixed, common elements to be interoperable (Galloway, 2004). In order to understand how this plays out with Wi-Fi, we need to start with what makes Wi-Fi wireless. Although the absence of licences and permits suggests an unruliness to Wi-Fi, important planning and legal arrangements underlie our ability to connect a Wi-Fi transmitter to a phone line without permission. The infrastructure that makes Wi-Fi wireless is not physical at all, but regulatory.

Spectrum as infrastructure

Wireless services work by transmitting radio waves as opposed to transmission carried across physical objects (fibre-optic or copper cables). Radio communication is achieved when an electronic device applies AC electric currents to an antenna. The result is the creation of oscillating electric and magnetic fields across the atmosphere, which can transmit information. A radio receiver is able to pick up these currents. Morse code, for instance, was developed to communicate long distances via electrical pulses along a telegraph wire, over radio waves, or even via a flashing light. When transmitted over radio, the sender produces code in the form of dits and dahs (represented as dots and dashes when written) by turning a current on and off with deliberate timing. With Morse code, the 'translation' of the signals into words occurs in the expert's head instead of through circuit boards (for instance, 'three dots, three dashes, three dots' is recognized internationally as a call for help). An early innovation included changes to the Marconi radio called frequency division (in 1900), which meant that a radio and receiver could operate on a specific frequency.

In the early days of radio, people who possessed the right equipment were able to transmit radio signals without needing permission. Military, corporate, and government interests at the international level eventually intervened. Their stated concern was safety at sea, as ships needed to receive signals for navigation but these were increasingly subject to interference. A radio signal is degraded if it experiences disturbance from an external source, including other signals.

The sinking of the *Titanic* in April 1912 is sometimes credited as the impetus for regulatory change, although in this case interference was not the reason the ship's distress signals went unheard. On 7 July 1912, the *San Francisco Call* called for the navy to take control of the broadcast spectrum to 'check wireless anarchy', writing that 'had the wireless been the perfect instrument for saving vessels in distress which it was held to be in the popular imagination, possibly every soul on board might have been rescued' (*San Francisco Call*, 1921, p. 22). However, as the article pointed out, the *Titanic* was equipped with one of the most advanced radios of its time, and its operator, who was certified by law, did send out distress signals. Unfortunately, the wireless operator on the Californian vessel 4 miles away was asleep in his bunk, and the signal went unnoticed. Spectrum interference was, however, said to be a critical issue that needed addressing, and the *Titanic* disaster gave visibility to the issue. By the end of 1912, radio operators and transmitter owners were required to obtain a licence (not just a certificate). Processes for designating spectrum for particular uses became law with the passage of the US Radio Act of 1927, and were coordinated at the international level by what was to become the International Telecommunications Union.

While transmitting data over the airwaves requires working with physical phenomena, what we think of as radio spectrum is not a state of nature. Rather, governments determined that they could exert power over the use of radio frequencies, assigning them for particular uses. Spectrum planning made sense in relation to the technologies that were around in the first half of the twentieth century as it resolved disputes related to signal interference. Radio and resulting infrastructures, including Wi-Fi, became embedded within the administrative technologies of state planning, licensing, and enforcement, which are themselves dynamic. Kevin Werbach, writing in 2004, observed that usable spectrum at that point in time was 5,000 times larger in terms of bandwidth than when the federal Radio Act was first adopted. The expansion was not a result of changes in the Earth's atmosphere but due to technological improvements that enabled devices to work more efficiently and at different frequencies.

Spectrum regulation involves the legal designation of frequencies for particular uses, and this in turn structures what behaviour occurs and in whose interest. In the US, the government initially allocated spectrum through central planning and comparative hearings, on the basis of maintaining social good with what was deemed a limited resource (carried out by the regulator, the Federal Communications Commission (FCC)). The result was a 'cozy oligarchy' (Marcus, 2009) of a small group of manufacturers who, unlike the computing industry, could slowly plan technological changes without fear of being overtaken by new entrants. However, from the 1950s, economists began to explore more dynamic models for spectrum allocation. Ronald Coase, for instance, proposed secondary markets in which those who hold a licence could bargain with others who wish to access part of that spectrum (Coase, 1959/2013). Underlying the argument for a market-driven approach to managing the airwaves was the contention that resource coordination could be achieved without creating interference through market mechanisms and property rights, as occurs with other resources. While spectrum auctions did eventuate in many countries – commencing with New Zealand in 1989 – spectrum is to this day regulated centrally and dealt with as a licence rather than a property right.

The FCC's decision to reserve part of the radio spectrum for unlicensed use was therefore a significant break with how things were done. The process began in the 1970s, motivated by a political and economic interest in deregulation (instigated by the FCC Chairman Charles Ferris rather than by industry, at the behest of the Carter Administration (Marcus, 2009)). In his personal account of this decision, Michael J. Marcus, who worked for the FCC at the time, describes it 'not as an attempt to bring specific products to market, but as part of a program to remove anachronistic technical regulations and allow a free market in innovation technology, subject only to responsible interference limits' (2009, p. 19). The final result, announced in 1985, permitted the FCC to authorize 'spread spectrum and other wideband emissions not presently provided for in the FCC Rules and Regulations' (FCC, 1984). Some parts of the FCC were unhappy with the outcome, leading to Marcus's transferal to another division and the disbandment of the team that oversaw the change.

The decision permitted the use of spread spectrum technologies in the police bands, as well as other unlicensed uses in spectrum that had been set aside specifically for industrial, scientific, and medical devices (ISM bands). Both components were important in the development of Wi-Fi: the determination to allow unlicensed use meant that Wi-Fi could be self-provided; spread spectrum referred to a particular technological design that would enable Wi-Fi and other devices to co-exist with less chance of interference. Today, spread spectrum technologies are widely used in low-power radio protocols. Every time a transmitter and receiver, detecting interference, jumps to a different frequency, it is using spread spectrum-related innovations.

Wi-Fi hardware and its precursors

When the FCC commenced the market assessment that prefigured the decision to open the ISM bands for unlicensed use in 1979, they did not have much of an idea of what these bands might be used for. Businesses were still using mainframes (Lemstra and Hayes, 2008), and wired local area networks (Ethernet) were unknown in most homes and offices. By the time the FCC process was completed in 1985, the internet had only just been named and the first portable computers were making it to market as luggable sewing machine-sized boxes with external battery packs (Hills, 2011).

Spread spectrum, however, was not a new idea. Although there is no indication that the FCC knew about it, the first patent for such a technology was lodged by Nikola Tesla on 17 March 1903 (Nordic Semiconductor, 2012). One of the most fascinating figures in technology history, Tesla worked on numerous inventions related to electric lighting, telecommunications, X-ray, radio remote-control, and alternating current (AC) induction motors. Despite producing hundreds of patents, his career suffered many setbacks, including an acrimonious departure from the Edison company, a failed start-up, and a fire that destroyed his writings, machines, and experiments. Tesla is today recognized as having commenced the development of numerous modern-day technologies, and possibly ones that may exist in the future. Tesla's experiments in wireless technology were directed at creating electrical power transmission without powerlines. He also invented wireless lighting, dazzling audiences with cordless gas-discharge lamps using his high-voltage, high-frequency alternating currents. Tesla never used the words 'spread spectrum', but described the synchronization of a transmitter and receiver as they hop between channels in a predetermined sequence, thereby avoiding interference.

During World War II, scientists and inventors on both sides of the Atlantic worked on solutions for pseudorandom radio systems that could transmit messages without being detected (meaning that they would appear random to an outside observer but could be detected by a receiver that was aligned to receive the signal as it hopped across different frequencies, and hence would avoid being jammed by the enemy) (Schwartz

Figure 2.1 Nikola Tesla holding a gas-filled phosphor-coated wireless light bulb *circa* mid-1896. *Source*: Tesla Universe.

and Abramson, 2009). Many of these techniques remained classified military secrets. Most famously, Hollywood star Hedy Lamarr and composer George Antheil filed US Patent no. 2,292,387, issued in 1942 for a secret military communication system for radio-controlled torpedoes, which would work according to a patterned sequence across eighty-eight carrier frequencies (also the number of keys on a piano). In the patent, they noted: 'Even if the enemy should pick up one of the impulses transmitted, he would not know whether it was an effective signal or a false signal. Furthermore, it is quite possible to so arrange the records that the receiver is never twice tuned to the same frequency' (Kiesler Markey and Antheil, 1941).

Despite a significant amount of work by numerous scientists and companies, these early experiments in spread spectrum did not lead to non-military innovations as spectrum rules would have rendered them unusable. Those working on the FCC rulemaking found there was little non-classified information on the technology (see Dixon, 1975). The scoping report commissioned by the FCC, known as the MITRE report (Scales, 1980), stated that spread spectrum technologies (frequency hopping, time hopping, direct sequence, and chirp) could be useful for interference-resistant technologies and that there 'are probably a number of potential non-Government applications of spread spectrum technology that have not received serious attention simply because designers of commercial equipment are generally not well-versed in this area' (p. 4). Later FCC deliberations referred only to the possibility of the vaguely named 'wireless data terminals' (Marcus, 2009), with no detail of what these might consist of. The decision to permit unlicensed use of particular spectrum bands is now considered remarkable, given that the technology that the FCC referred to in its inquiry was not much more than an 'esoteric topic' (2009). The FCC's Marcus recalls: 'The spread spectrum goal at the time was not to introduce a specific class of products, such as wireless local area nets, or even a specific band, but rather to create relatively clear opportunities for this technology to reach market in order to encourage investment in R&D' (p. 21).

However, other developments in wireless technology were occurring. One notable precursor to Wi-Fi was the ALOHAnet, developed by the University of Hawaii in 1971. Engineers at the university set out to use radio communication to share resources (conceived at the time as the computing abilities of hardware) with campuses on other islands, recognizing that radio presented 'design options not available in systems using conventional point-to-point telephone channels' (Schwartz and Abramson, 2009, p. 21). As there was no unlicensed spectrum available at the time, ALOHAnet used two temporary experimental 100 kHz assigned channels, which allowed for a range of within 100 km. The system was designed to send user information in a single high-speed packet burst in a shared channel, known as an ALOHA channel (Schwartz and Abramson, 2009). The innovation of ALOHA was 'random access' (now known as 'direct access') – a computing data structure that allows for data to be sent in any sequence, rather than requiring them to be ordered sequentially, which is less efficient. This in turn led to the development of carrier sense multiple access (CSMA) and CSMA with collision avoidance (CSMA-CA) used in Wi-Fi (whereby algorithms verify the absence of traffic before transmitting across a shared medium, transmitting when a channel is idle).

In the early 1970s, a grad student at Harvard turned his attention to ALOHA and made it the subject of his second attempt at a Ph.D. thesis (his first attempt, on ARPANET, was not accepted). The thesis highlighted the significance of the point-to-point circuit-switched and packet-switched data networks along with random access. The student was Bob Metcalfe, who would later go on to pioneer cable networking technology with David Boggs at the Xerox Palo Alto Research Center, using random access. That first version of Ethernet

Figure 2.2 ALOHA terminal control unit, 1971. *Source*: Schwartz and Abramson (2009, p. 22).

(1973) was originally called the Alto ALOHA Network and formed the basis of the open standard for Ethernet, which Metcalfe commercialized after leaving Xerox. Although ALOHA and Ethernet laid the foundations for the development of Wi-Fi, Metcalfe himself didn't see a future in wireless. In a 1993 article he wrote: 'Simply put, there aren't enough megahertz to go around out there in our increasingly polluted electromagnetic ether. It is an ecologically unsound waste of energy to broadcast bits in all directions when they need to be received in only one. The ether is too scarce to be wasted on nonbroadcast communications, and it won't be' (Metcalfe, 1993, p. 48). As Wi-Fi had no future, he suggested that everyone 'wire up our homes and stay there' (p. 48). This was not the only time that Metcalfe's predictions would be wrong. He was forced to eat his words after writing that the internet would collapse in 1996. Metcalf took a print-out of the article in which he made the statement, blended it with some liquid, and drank it at the International World Wide Web Conference in 1997.

The use of cellular/mobile telephony infrastructure for transmitting data (including SMS and emails) developed in parallel with the wireless local area networks that would become Wi-Fi. Cellular Digital Packet Data was launched by Bell Atlantic Mobile in the US, initially available in Washington, Baltimore, and Pittsburgh in 1994. The system was described by the *New York Times*: 'Instead of dialling up a number and securing a full cellular channel to send a message, the new system sends each message as a series of electronic envelopes or "packets", each of which contains an address for its destination. These packets are slipped in between the voice signals of ordinary cellular telephone users' (Andrews, 1994). Each message would cost customers a few cents, which was 'far cheaper' than a minute-long telephone call. The service did not gain market traction and was taken over by the faster General Packet Radio Service (GPRS) technology, and eventually shut down in 2004.

Of all of these efforts, it was arguably cash registers that ended up directly informing the development of Wi-Fi as we know it. NCR Systems Engineering, a Dutch company formerly known as National Cash Register, began working on a solution that would enable electronic cash registers to connect to each other without cables (making store layout changes cheaper and easier). As they were interested in selling cash registers to the US, the company needed to determine what rules their products would have to conform to in order to gain certification from the FCC. Through the process of determining the most appropriate technologies to incorporate into their wireless local area network development, individuals at NCR became involved in the standards processes, eventually leading a new working group to develop what would become the Wi-Fi standard: IEEE 802.11.

Wi-Fi standards

Standards are orderings that enable us to recognize when something is 'the same as' for the purposes at hand (Busch,

2011). They are 'boundary objects' in that they allow people and things to work together without consensus (Star, 2010). Users of Wi-Fi might not understand the technical arrangements that enable connectivity, but they understand (or can learn) that Wi-Fi is a particular means of connecting to the internet or to other devices when an icon is displayed. In this way, boundary objects are a shared structure, which can be physical or conceptual (for instance, a library catalogue or a theory), where there is delineation over what something is or is not. A boundary object scales up to become infrastructure when it becomes standardized through sets of rules and determinations that formalize how something is to work.

The power of a standard rests in 'the ability to set the rules that others must follow, or to set the range of categories from which they may choose' (Busch, 2011, p. 29). Busch discusses this in relation to Heidegger's notion of handiness – that standards make things handy, and handy for all in the same way. The significance of this in relation to Wi-Fi is that an ecology of devices all uses the standard, so that when you switch on a device you can be confident it will connect. Standards thereby stabilize a set of practices and enable otherwise separate manufacturers to create things that work together.

The story of the development of the Wi-Fi IEEE 802.11 suite of standards is detailed in a book by technology policy academic Walter Lemstra with the first chairman of the IEEE 802.11 working group, Vic Hayes, and economist John Groenewegen (Lemstra, Hayes, and Groenewegen, 2011). They tell of how individuals from NCR, including Hayes, undertook to lead a process for the development of a standard for wireless local area networks (LANs). The process also demonstrates something of the nature of Wi-Fi as an infrastructure. In relation to the dimensions of infrastructures outlined in Star and Ruhleder (1996), the history of Wi-Fi standards shows how communities of practice play a role in the formation of infrastructures, and that they can enable and limit development.

The IEEE is a professional association based in New York, but with members from around the world. Aside from research publishing and education, the association also participates in the creation of standards, including the standards that apply to Wi-Fi and related technologies. The first meeting of the 802.11 working group took place in September 1990, after it was decided that the existing committees for wired LAN were insufficient for the task at hand. Wi-Fi was nonetheless developed to work with the standard for wired local area networks, and therefore contains standards that are shared with Ethernet. As is often the case, multiple standardizations can occur within an infrastructure, reinforcing each other in a network of nested relationships (Lampland and Star, 2009).

An early and important decision was the commitment to open standards. An open standard is one that is developed in a transparent manner, free or licensed on a reasonable and non-discriminatory basis, driven by stakeholders, publicly available (not necessarily free of charge) and maintained (ICT Standard Board definition in Lemstra, Hayes, and Groenewegen, 2011). NCR and its allies Symbol Technologies and Xircom pushed for an open standard, having been held back in earlier R&D efforts by existing proprietary IBM protocols used for connecting mainframes to mini computers. The resulting distributed control architecture (as opposed to IBM's central architecture) enabled what is known as Wi-Fi ad-hoc networking. Connecting game consoles or other consumer electronic devices are common examples of ad hoc Wi-Fi transmission, as is tethering devices in order to share internet connection, thereby creating virtual routers. Mesh networks, discussed further in Chapter 6, are also possible because of this decision. However, some home and office mesh networking systems (used to extend Wi-Fi signals) have been developed with other proprietary features, which means they do not connect to other brands' devices. In 2018, the Wi-Fi Alliance released the Wi-Fi EasyMesh standard in order to encourage the industry to produce devices that could extend networks regardless of brand. Other significant battles described in that history include the decision to develop two standards for two different spread spectrum modulation techniques – direct sequence and frequency hopping (IEEE 802.11a/b) – and how this resulted in interoperability problems between devices in the early days. These issues arose primarily as a result of competing interests, whereby individuals representing their employers would campaign for particular technologies to be included in the standard. Despite this, the history by Lemstra, Hayes, and Groenewegen (2011) demonstrates multiple instances of compromise being achieved in order to reach mutually beneficial outcomes. Even though the people involved tended to be aligned with a company, they often acted in the interests of the industry as a whole, motivated by a passion for technology and casting votes in favour of technical merit.

One of the problems with standards is that they can keep things the same for longer than is desirable: standards 'make it costly (in terms of money, skill, organization, and social networks) to shift to an alternative development path since future actions are contingent on those taken in the past' (Lampland and Star, 2009, p. 61). The IEEE uses formal decision-making processes, themselves being standards that need to be followed. Each time a Wi-Fi standard changes, it goes through a series of formal motions, a period for comment and debate, and a ballot. The dynamics and governance of the group during the development of IEEE 802.11 therefore required investments of time and material resources from those involved or their employers. These requirements can create path dependence, making it difficult to shift direction. In this way, standards encourage the investment of time and material resources.

While debates within the IEEE were occurring, different products were being developed and sold, competing on price and speed. One of the first civil applications of spread spectrum was for local area networks used for cordless microphones in music concerts in the 1980s (Lemstra and Hayes, 2008; Hawkins, 2017). NCR released its first WaveLAN product to the market in 1991. Lemstra and Hayes (2008) described it as follows: 'The product operated at 915 MHz and used one communication channel providing a bandwidth of 2 Mbit/s. It was a desktop PC plug-in board, essentially a radio-based Network Interface Card (NIC), and required an external antenna' (p. 51).

However, NCR's WaveLAN was expensive, and multiple access points were needed to cover a building. At Carnegie Mellon University, a research network pioneered methods for covering large areas using wireless LAN. Named after the founders of the institutions that eventually became Carnegie Mellon (Andrew Carnegie and Andrew Mellon), the Andrew Project provided coverage to the campus (Wireless Andrew) and led wireless research in the areas of security, network management, integration with cellular networks, and policies for the management and operation of wireless networks (Hills, 2011). The work of the Carnegie Mellon researchers, including Alex Hills, grappled with the ways in which infrastructures are 'built on an installed base' (Star and Ruhleder, 1996, p. 113), and overlaid onto other infrastructures in ways that can be convenient or limiting.

At the time when these experiments were occurring, there was no guarantee that Wi-Fi would gain market traction. A significant breakthrough came in the late 1990s when Steve Jobs decided that Apple would include a wireless network interface in its new iBook laptop (the 'AirPort'). The NCR team, which had by then been acquired by Lucent Technologies, set about producing a low-cost version. In 1999, they achieved a card priced at $99, with an access point for $299. Dell followed suit, with wireless available on PC computers not long after Apple's release.

In that same year, the Wireless Ethernet Compatibility Alliance (WECA) was formed, in part to deal with the problem

Figure 2.3 Apple AirPort base station. © Mark Richards. *Source*: The Computer History Museum.

of the two competing spread spectrum variants, frequency hopping and direct sequence. WECA demonstrates both the possibilities and the limits of technical standards: a certification process was required in addition to the standard, because the complexity of the standard was such that it did not ensure compatibility. Companies could fully comply with the standard yet still produce products that would not work together. WECA drove the adoption of the DSSS-based standard (IEEE 802.11b) and established a testing procedure and a seal of compliance. Companies wishing to display the trademarked Wi-Fi logo would need approval from WECA, which changed its name to the Wi-Fi Alliance in 2002 (Lemstra and Hayes, 2008).

Since the first version of Wi-Fi was approved, a number of technologies have been added to it, while others have been declared superseded (see appendix 3 of Lemstra, Hayes, and Groenewegen, 2011). The additions have enabled more data to travel within the frequencies that devices operate in, changed the techniques used for generating a signal, added security features, enabled the compatibility of Wi-Fi devices with other devices, expanded the number of transmitters and receivers that are able to operate simultaneously, and directed transmission to specific devices. The result over time has been longer-range transmission, faster speeds, and less vulnerability to hackers. We discuss home networks further in Chapter 3, including how wireless gateways – the component of Wi-Fi infrastructure that enables data to enter and exit the home – has intensified the home as a site of information and entertainment consumption.

Wi-Fi has also developed from short range for connection between devices, to an augmentation infrastructure that other technologies use to offset capacity, to connect large-scale networks of things, and for the collection of data. In 2018, the IEEE approved a new standard, IEEE 802.11ax, which the Wi-Fi Alliance has rebranded as Wi-Fi 6. In the process, it also renamed the previous two standards Wi-Fi 5 (802.11ac) and Wi-Fi 4 (802.11n), finally relegating the clumsy 'IEEE 802.11 and its variations' to history. Over the course of its evolution, Wi-Fi has become increasingly embedded in other infrastructures. It is common for users to switch between Wi-Fi and mobile data, seeing mobile as a direct connection and Wi-Fi as a borrowed or shared connection. While this assumption is true in relation to our data plans, at the technical level it is now much more complex. Mobile calls on some networks divert to Wi-Fi as a means to conserve capacity. Recent versions of the internet's Transmission Control Protocol are designed to facilitate ready switching between access modes, such as cellular and Wi-Fi, without disrupting data flows. Smartphones now have Wi-Fi calling capabilities, which some telephone companies will charge as a standard call even though the connection itself is free. In addition, 5G and Wi-Fi 6 are interoperable in both directions, meaning you might access someone's Wi-Fi but it connects you to their 5G connection.

On the regulatory side of this story, policymakers in many countries are now working to release more spectrum for non-traditional 'shared' use, whereby devices are not restricted to a particular frequency, but able to move across frequencies and dynamically fill vacant spectrum. These moves recognize that technologies have evolved so that signal interference can be avoided. Thanks to the FCC's 1985 spread spectrum decision, devices now coordinate to make sure that the transmitter and receiver switch to a different frequency to avoid other signals. Increasing demands on spectrum mean that governments are now looking at releasing more spectrum for unlicensed use by these so-called 'smart' devices.

Wi-Fi infrastructure and development

Wi-Fi can be an aspirational infrastructure (Larkin, 2013), in that it can be used to provide internet to those who have none, particularly where other internet infrastructures are not considered viable. The aspiration or expectation is that Wi-Fi can overcome digital exclusion, providing people with opportunities and services online. In the final section of this chapter, we consider the political nature of Wi-Fi when it is provided to address digital exclusion. The two cases we use – remote Australia and remote Malaysia – also illustrate how Wi-Fi as an infrastructure differs in relation to, or in the absence of, other infrastructures. Households in both of these remote locations are far less likely to have access to the internet than households in larger towns and cities in their respective countries.

Infrastructures provide us with comfort and convenience, shaping aspects of the natural world into a form that we find useful and overcoming features that we find difficult. As Paul N. Edwards writes, 'infrastructures are largely responsible for the sense of stability of life in the developed world, the feeling that things work, and will go on working, without the need for thought or action on the part of users beyond paying the monthly bills' (Edwards, 2003, p. 187). Many studies of infrastructures focus on a particular type of infrastructure and look at the practices they enable. However, infrastructures can also be studied in relation to how they co-construct modernity, providing a linking role in social organization and forming 'the stable foundation of the modern social worlds'

(p. 186). In relation to communication infrastructures, this includes the expectation that we can always be connected to the internet, and that it will work at a tolerable speed. The 'modernity' of internet infrastructure is one in which users no longer need to wait or contend with weak signals, or find themselves without connectivity in the places where they live and work.

The macro-level infrastructural dimensions of Wi-Fi become apparent when other telecommunications infrastructures are not present, such as when Wi-Fi is deployed in remote areas as a strategy for digital inclusion. As other internet infrastructures improve, those who have limited or no access to the internet find it increasingly difficult to perform certain tasks. Many services today, such as banking and employment services, are designed to be accessed 'digital first', meaning that citizens are expected to use digital platforms when making contact. For those of us who have adequate internet, this becomes the stable and convenient environment we expect. However, those without access can miss out on essential services, or bear additional costs to gain the same benefit (such as bank fees for branch transactions, or transport costs to visit an employment agency).

Some of Australia's remote Indigenous communities are vast distances from government services and retail stores. The Aboriginal and Torres Strait Islander families who live in these communities choose to do so in order to maintain connections to their traditional lands and continue cultural practices. These communities are often only reached by unsealed roads, which can become inaccessible from floods in the wet season, while, in the drier months, temperatures in the interior can soar above 50 degrees Celsius.

Online services should make life easier for those living in remote areas. However, remote locations have always posed particular problems in relation to telecommunications infrastructure. They can defy telecommunications markets, in that building and maintaining infrastructure for a small number of people means that the original cost is not likely to be recovered through standard charges. Geographic obstacles can make the installation and maintenance of infrastructure difficult and expensive. Cultural and economic systems can also play a part. For instance, residents in Australia's remote Indigenous communities may apply for a subsidy to have a satellite dish installed on their roof in order to receive a home internet connection. However, this option has been unpopular among Aboriginal households as there is no pre-paid option for satellite internet, and post-paid billing can be difficult to manage in an economy characterized by cash transactions and shared resources (Rennie et al., 2016).

The 2016 Census of Population and Housing revealed that, within the Aboriginal and Torres Strait Islander population, there are significant differences based on location: 82.8% in major metropolitan areas access the internet, compared with 73.2% in regional areas, 61.3% in remote areas, and 49.9% in very remote areas. However, when we drill down to the infrastructure level, it becomes clear that some communities are much worse off than others. An infrastructure survey of 401 small Indigenous communities in the Northern Territory, conducted in 2016 by the Centre for Appropriate Technology (CAT), found that only 20% had mobile phone coverage (CAT, 2016). Moreover, while some communities were documented as having internet connections, this was often extremely limited. Of the 140 communities that had an internet connection of some description, in 80% of these the internet was only available at one house: 'One community in the Utopia Homelands, for example, is described as having internet access; however, the community has a population of more than 40 people and it is only the three people living in the one house with internet access that can use the internet' (2016, p. 42).

In order to address the problem of digital exclusion, the Australian government has funded programmes that provide Wi-Fi or computer centres in remote Indigenous communities where commercial telecommunications services have not reached. Wi-Fi is a practical approach in that it provides the ability to share a satellite connection, overcoming the need for cabling. Installation of a satellite dish in a remote area is cheap compared to the costs of installing fibre-optic cables. In addition, Wi-Fi hotspots in remote locations such as these are generally powered by small solar panels, which overcomes the need for electricity grids or generators.

Between 2013 and 2015, a telecommunications company (Australian Private Networks) that was administering satellite public phones to communities of under fifty residents undertook to upgrade the phones to provide Wi-Fi. During this timeframe, APN installed Wi-Fi into 301 communities. While these services are valued in communities, they provide limited internet access compared with that which is available in other parts of Australia. Using standard equipment, Wi-Fi will likely only reach houses in the immediate vicinity of the phone box, with signals obstructed by walls and other physical objects. The download limit for each satellite backhaul is capped, meaning that a limited amount of data is shared between all users of the Wi-Fi network. Community providers typically limit the amount of data each user has access to per day in order to avoid the data allowance being used up at the start of the month, which restricts the activities that people can do online. While Wi-Fi companies have experimented with pay-per-use models, they have encountered low use of these payment systems in some localities, particularly if residents are without credit cards and where there is no local store to sell vouchers (Rennie, Yunkaporta, and Holcombe-James, 2019).

The interior of the Malaysian State of Sarawak is physically very different from Australia. Located on the island of Borneo, Sarawak is covered with dense rainforests, mountains, and palm oil plantations. In the Baram region, small villages (*kampung*) are dotted along a network of rivers, which act as transport corridors for longboats. The road system in this part of Sarawak is privately constructed and maintained by logging companies. Roads can become unusable depending on the operations of the logging companies at the time, and are prone to mudslides from the surrounding mountains and the jungle's fervent vegetation. Like Australia, government programmes have looked to Wi-Fi as a means of providing internet to those living there.

In Sarawak, the Malaysian government opted to install small-cell mobile telephone towers instead of providing a subsidy for satellite internet. A survey of infrastructure in eleven communities in the Baram region of Sarawak conducted in 2016–17 found that only 36 per cent of people were using the internet at the time (Horn and Rennie, 2018). Of these, 70 per cent were using the internet daily in their village, with the percentage of daily users rising by 20 per cent when usage in other places was included (in the city, or towns with decent access). Seven of the eleven villages had mobile coverage, but for six of these this was small-cell coverage requiring people to stand within 150 metres of a tower to receive a signal.

In some situations, funding for the towers was provided in the expectation that companies would install mobile telephone equipment, but this did not occur (Horn et al., 2018). Some communities that did not receive small-cell infrastructure were instead provided with public Wi-Fi hotspots, typically installed outside a local shop or in a communal area. Wi-Fi was also provided to schools for teaching and learning purposes. However, during interviews, residents told of how the public hotspots would be slow, with extremely limited data. School teachers said that they were often unable to use the school wireless for their own work, let alone share it with the community. The maintenance of Wi-Fi hotspots was also a problem for communities, as residents had low technical skills.

Wi-Fi does provide a basic level of internet access, enabling people to use low-bandwidth services that they might otherwise go without. However, in both contexts, the presence of Wi-Fi masked the dynamics of digital exclusion, such as slow speeds, inadequate data limits shared between a large group of people, and the physical reach of Wi-Fi networks. Policy reports can show that communities have connectivity when, in fact, networks are limited or poorly maintained. In this way, Wi-Fi is an aspirational infrastructure in that it promises a low-cost alternative to other internet infrastructures, leading to expectations that it can assist people to experience the opportunities of the internet, such as online services, shopping, and education. When viewed close-up, Wi-Fi that has been provided for such purposes may provide a basic level of internet service, yet can be vastly inferior to other internet infrastructures.

Conclusion

Wi-Fi has become a common infrastructure in a relatively short period of time, partly because it can be easily assembled with only basic technical skills. The story of Wi-Fi's creation demonstrates that this was achieved through spectrum design and policymaking, the careful work of standard setting within the IEEE, and the innovations of various hardware and software components by multiple actors. As an infrastructure, Wi-Fi extends the means by which the internet is accessed, used, and shared. It has been promoted within policies and programmes aimed at overcoming digital exclusion precisely because it is shareable and low cost. Wi-Fi can work in the absence of other on-the-ground infrastructures, requiring only a satellite dish for backhaul and solar panels for power. At the same time, the evolution of Wi-Fi has seen it increasingly entwined with other infrastructures, including mobile networks, geolocation technologies, and the internet of things. It will be important to gauge the extent to which the benefits of Wi-Fi differ in areas where these other infrastructures exist, compared with those places where it is the only point of access.

3
Home

There is a joke about household Wi-Fi use that runs something like this:

Q. What is a foolproof way of calling an impromptu house meeting?

A. Turn off the Wi-Fi router and wait in the room in which it is located.

As jokes go, this one isn't particularly funny. What this joke does do, however, is usefully convey three things that are central to the concerns of this chapter. First, it highlights how Wi-Fi has come to be regarded as a key domestic technology, one that is now seen as integral to and at the centre of contemporary family life. Second, it shows that Wi-Fi serves as a vital gateway (more on this below) through which other forms of information, communication, and media content flow. And third, it reveals that Wi-Fi impacts established household dynamics in manifold and significant ways.

'The advent of Wi-Fi', Taylor, Middleton, and Goodrick (2015, p. 48) note, 'was enabled by a leap of faith by the US Federal Communications Commission (FCC) in the 1980s to encourage use of then-unknown new wireless technologies that did not require exclusive licenses. Wi-Fi use exploded and it has now become a ubiquitous household technology based on shared spectrum principles.' Indeed, as we have noted, the number of Wi-Fi-enabled devices globally in 2020 is believed to exceed 13 billion (Wi-Fi Alliance, 2019b), up from 4 billion in 2010 (Mackenzie, 2010).

Growth in Wi-Fi-enabled devices is nowhere better reflected than in the domestic home. As Gerard Goggin has pointed out, 'while Wi-Fi was largely pioneered by researchers and 'early adopters' in commercial or corporate settings, ... diffusion to wider populations has [largely] come through its non-commercial use in households' (Goggin, 2007, p. 121). The domestic home has thus both served as an important site for facilitating the 'national domestication' of Wi-Fi (p. 121) *and* continues to constitute a crucial site in its own right, especially for the consumption of wired and wireless broadband services. In the US context, Wi-Fi is by far the most common way that households access the internet (Hamilton, 2017). Wi-Fi-enabled US households are said to own 'an average of 5.7 computing devices and 8.1 connectable CE [consumer electronics] devices' (2017), and almost 98% of broadband data use occurs via Wi-Fi (MarketWatch, 2014). While other solutions for household connectivity have been developed, including Bluetooth, Zigbee, and WiMax, among others, Wi-Fi has maintained its popularity for a range of reasons, including, at least in part, due to its low cost, and because of the way that it operates 'out of control', as Adrian Mackenzie (2010, p. 4) puts it. By this, Mackenzie means that Wi-Fi 'requires little centralized infrastructural management' (p. 4) in order to operate successfully. The result is that Wi-Fi now 'rules the home' (Communications Technology, 2009).

Wi-Fi works by channeling data through 'wireless gateways'. A wireless gateway routes packets of data from a wireless local area network (LAN) to another network, which could be a wired or a wireless wide area network (WAN); a wireless gateway also combines the functions of a wireless access point, a router, and, often, a firewall as well (Wikipedia, 2017). A 'residential gateway' is effectively a scaled-down version of this, in which a small, consumer-grade router provides network access between LAN hosts to WANs via cabling and a modem (Wikipedia, 2019). Typically, the cable broadband service connects to the router, which then makes broadband internet available to a range of Wi-Fi-enabled devices, such as laptops, tablets, and smartphones. It is literally through this residential gateway – the Wi-Fi router – that data enters and exits the home. The proliferation of devices within the home, and intensified consumption of cross-media diets (Sandvik, Thorhauge, and Valtysson, 2016; Haddon, 2016, p. 25), are enabled and sustained by Wi-Fi. Indeed, as a 2016 Ericsson Mobility Report notes, 'over 85 per cent of data traffic generated by the use of smart phone video apps goes over Wi-Fi' (cited in Telecomlead, 2016), and, although data usage on smartphones is growing, Wi-Fi data is dramatically outpacing it. This points to why it is that access to, and control over, the residential gateway of the Wi-Fi router can be so hotly contested. It is around and through this device that data access, within the contemporary home, is negotiated within households. In this sense, Wi-Fi forms an important domestic boundary object (Star and Griesemer, 1989) in that it is simultaneously working to meet the needs of those within the home, while also subject to various forms of constraint. It is, as we shall argue in this chapter, both a literal gateway through which data enters and exits the home, and a vital figurative gateway around which access (and for whom) is controlled, contested, and negotiated.

How are we to make sense of the significant role that Wi-Fi plays as a gateway technology within the home? Within media and material culture studies, the domestication approach has come to form a key critical lens through which to view the entry and settlement of information and communication technologies within the home. In this chapter, we adopt this framework to examine how household Wi-Fi has been domesticated. The domestication framework has been applied to wider studies of Wi-Fi provision and use, including in relation to municipal free Wi-Fi in disadvantaged communities (Chigona et al., 2016) and in relation to work-related Wi-Fi use within the public–private space of cafés (Henriksen and Tjora, 2018). Yet, curiously, Wi-Fi rarely figures (at least not explicitly) in studies of household technology use that employ the domestication framework. A key argument of this chapter is that Wi-Fi is not only increasingly central to our everyday engagements with media and information and communication technologies within the home, it in fact plays a key role in *animating* the home as a centre of media and technology consumption. In vital ways, Wi-Fi has changed the way that households work – Wi-Fi has *domesticated* the household – and it has transformed the household media economy. Wi-Fi frees us from dependency on single devices (such as the single desktop computer or loungeroom television), it permits device interoperability and mobility within the home, and it works to make the home more 'social'. What will become clear from the ensuing analysis is that Wi-Fi does not always necessarily map neatly onto the domestication framework as classically understood. Wi-Fi also prompts us to rethink traditional domestication approaches, due to the ways that it draws attention to the recursive qualities of technology.

Nevertheless, the domestication framework remains a useful critical framework with which to begin when setting out to understand the importance of home Wi-Fi, especially for household technology consumption and use, and in making sense of the interplay between the household and the outside world. Indeed, far from invalidating the domestication framework, or marking it as outdated, we view these points of fit and misfit as a productive tension, a valuable means of understanding the place and importance of Wi-Fi within, and to, the home. This perspective also feeds into one of the larger arguments of this book – that Wi-Fi 'displays constant contractions and dilations, and multiple instantiations', and how, because of this, and its 'plurality and lack of coherence', Wi-Fi poses certain challenges for media studies and established media methods (Mackenzie, 2010, p. 13). From this investigation of the domestication of Wi-Fi, in the final section of the chapter we trace the continued importance of Wi-Fi to the connected home, and provide a number of examples of how Wi-Fi technologies and standards are continuing to evolve in order to respond to the changing demands and patterns of household technology use, and the proliferation of connected devices within the home.

The domestication of Wi-Fi

Since the turn of the last century, households have undergone a series of transformations concerning the number, variety, and intensity of digital media and communication technologies, and an associated transformation in the way people use and interact with them. 'Our domestic life', Roger Silverstone and Eric Hirsch (1992, p. 1) write, is 'suffused by technology, and information and communication technologies are becoming a central component of family and household culture'. The domestication approach was developed as a way of making critical sense of this technological profusion and suffusion within the home.

The domestication approach, which emerged from material anthropology and the sociology of consumption and has since been adapted and applied in media and technology studies, provides a framework for describing and analysing the processes by which technologies are integrated into, and domesticated – or 'tamed' (Haddon, 2011, p. 312) – within, everyday life, and the ways that users and environments change and adapt accordingly.

Within the domestication literature, the home is regarded as a privileged site (Haddon, 2007) for two key reasons. First, because it forms a key, largely private, site into which new consumer products and goods – especially media, and information and communication services – are brought and consumed. Second, because it is here, within the domestic home, that these technologies are appropriated, incorporated, and 'redefined in different terms, in accordance with the household's own values and interests' (Silverstone, Hirsch, and Morley, 1994, p. 16). That is to say that it is here that 'the public meanings inscribed by and through [these] commodities, beliefs, and media and information consumption are ... open to negotiation' (p. 17). These negotiations, these authors argue, are defined by and filtered through what they refer to as the 'moral economy of the household' (p. 17; see also Strain, 2003).

While a number of previous domestication studies have focused on single technologies, like the computer (e.g. Lally, 2002; Haddon, 1994), this single focus is increasingly regarded as both undesirable and untenable as a result of the proliferation of mobile, streaming, and IoT (internet of things) devices entering the home. There have, for instance, been calls for a more holistic approach that considers domestic information and communication technologies – following Bausinger (1984, p. 349) – as 'media ensembles' (Haddon, 2006, p. 195) that are entwined within 'complex media repertoires' (Haddon, 2016), or as forming complicated media ecologies (Nansen et al., 2011; Shepherd et al., 2007). While we are very sympathetic to these arguments (e.g. Wilken et al., 2014; Armitage et al., 2017), we persist here with a focused examination of the domestication of a single technological form – Wi-Fi – for reasons that sit at the very core of the claims of this chapter: that Wi-Fi animates the home, and serves as a key (yet generally neglected) enabler of the 'cross-media diets' (Haddon, 2016, p. 25) that are consumed within, and the complex media repertoires that saturate, the home.

The domestication framework, as Silverstone (1994) and Silverstone, Hirsch, and Morley (1994) conceive of it, is built around five interconnected processes, or five 'elements of the transactional process' (1994, p. 21): *imagination* (wherein commodities are constructed as objects of desire within advertising and marketing systems of meaning creation, prior to purchase); *appropriation* (the process of purchasing a commodity and bringing it into the home); *objectification* (the process of physically and symbolically placing objects within the home); *incorporation* (processes of technology use within and near the home); and *conversion* (the process of defining the relationship between the home and the outside world). In the sections that follow, we consider Wi-Fi in relation to each of these five elements.

Imagination

In his book *Television and Everyday Life*, Roger Silverstone makes the point that, 'in the actual practice of consumption, goods are *imagined* before they are purchased' (Silverstone, 1994, p. 125 – emphasis added; see also Silverstone and Haddon, 1996). This forms a key, preliminary phase in the domestication process, in which householders speculate about what it might be like to have particular goods or services in advance of acquiring them. Print and online media play a vital role here by disseminating information about particular goods and by providing a forum for advertisers and marketers who seek to propagate ideas about these goods and to circulate idealized visions of what they are or might be. With Wi-Fi, it is possible to gain access to and understanding of the imagination phase by examining early journalistic explanations of how it works, and advertisements for Wi-Fi routers. Here we draw on a few examples from a search of catalogues from US store Radio Shack, and from the Internet Archive's online repository of computer and consumer electronics magazines for the period spanning 1997 to 2004, the first seven years of Wi-Fi's existence as a consumer product and service. While many of these publications tend to cater to the technologically informed (including gamers and computer enthusiasts), they are nonetheless instructive for grasping how Wi-Fi was positioned historically as a consumer product, and how consumers were likely to imagine it being appropriated and incorporated into their homes.

What is striking about this archival research is that the first mentions of Wi-Fi in these publications did not appear until 2001, four years after its initial consumer release.

Even so, both at and prior to this point, Radio Shack's catalogues featured innumerable articles on and advertisements for modems (an explainer on computer modems was published in the 1997 edition of their catalogue), mobile devices (ranging from walkie-talkies, cordless phones, and hands-free phone headsets, to portable music devices, and, later, digital mobile phones), and wireless systems. The last in this list included 2.4 GHz 'room-to-room wireless links' for distributing televisual content to a TV in another room (Radio Shack, 2001, p. 174), and wireless arrangements for dispersed household audio systems in order to achieve 'total listening freedom' (Radio Shack, 2003–4, p. 39). The promotion of these devices serves an important precursory and anticipatory function insofar as they work to prepare households who consume Radio Shack's catalogues for the arrival of Wi-Fi by acclimatizing them to the idea of networked devices, internal household mobility, and 'wirelessness'.

In the review section of the May 2001 issue of gaming magazine *PC Powerplay*, there is a full-page article introducing readers to Wireless LAN, a technology that is described as 'very much in its infancy' (*PC Powerplay*, 2001). While heavy on technical detail, the article concludes that 'for gaming the possibilities are endless: not only could you play games with other people within your house but, conceivably, a friend from across the road' (2001). By 2003, however, there is an assumption that committed PC enthusiasts and gamers would already have Wi-Fi. In a 2003 issue of *Atomic* magazine, there is an extended feature on Wi-Fi's technical aspects, including detailed coverage of its perceived limitations (signal loss, especially over distance), with DIY instructions on how to build signal-boosting Wi-Fi antennas (Chia, 2003).

Unlike the aforementioned specialist computing magazines, the 2003–4 issue of Radio Shack's 'Everyday Needs' reference guide, a reincarnation of their long-running yearly catalogue, is the first time that their publications carry explicit mention of Wi-Fi. In this guide there appeared a two-page spread selling the virtues of wired networking (on one side of the page) and 'wireless networking for home or small office' (on the other side of the page) (Radio Shack, 2003–4). In the page on wireless networking, the following scenario, depicting competing computing demands, is presented: 'You've probably experienced the household conflict for Internet availability. Mom wants to research the family vacation. Sis wants to get online and chat. Sonny needs to print his homework, and Dad wants to check stock prices. Now you can do it all at once with instant Wireless Networking' (p. 67).

Accompanying this text is an image of a man, sitting at a desktop computer, sharing a laugh with a woman sitting beside him, who has her legs extended on a recliner and a large portable computer open on her lap. Wireless networking, the reader is told, permits household mobility ('imagine having the flexibility to access your network from just about anywhere in the house'), connectivity of multiple devices and the streamlining of space ('you can link all of your computers together to share resources without having to deal with cables and wires'), and system flexibility ('wireless networking products ... give you the power to set up your network your way') (p. 67). Below this is 'Q&A' text explaining two variations on the 802.11 wireless protocol for readers, and, below that, a series of pictorial ads for various household wired and wireless Linksys devices, including a 'Linksys wireless router/access point and 4-point switch' (p. 67).

In 2004, consumer electronics magazine *CE Tips* published a multi-article feature section extolling the virtues of 'cutting the wires' and installing a wireless home network (Baker, 2004). The five articles in this feature section move from an introduction to Wi-Fi ('Wi-Fi 101') – which includes basic information on how Wi-Fi works, as well as a 'Wireless Networking Dictionary' for readers – through to instructions for the more adventurous consumer on setting up a wirelessly connected home entertainment and gaming system. As with the Radio Shack article, the key imagined benefits for the householder include increased internal mobility, connectivity of multiple devices, and the streamlining of space ('who wants nasty wires strewn across his home?', p. 38).

What is also important to remember is that the process of householders drawing on the media in order to *imagine* Wi-Fi as both product and service is an ongoing one. Throughout the 2000s, for instance, computer and consumer electronics publications like those mentioned above were dotted with pictorial ads for Wi-Fi routers, touting benefits such as 'powerful home networking performance', 'optimiz[ation] for entertainment', and 'easy Internet access for guests' (a 2010 ad for (then Cisco-owned) Linksys E series routers – see Radio Shack, 2010).

Jumping forward by almost two decades, we can also consider what is being imagined with the arrival of Google's domestic Wi-Fi offering: 'Google Wifi'. In one video commercial pitching Google Wifi to domestic consumers, the viewer is presented with 'a new connected system for fast Wi-Fi throughout your home' (NovaTech SE, 2016). This system consists of three squat, white cylinders of 'subtle design', with a pale blue band around their mid-section; these cylinders are known as Wi-Fi 'points'. The commercial shows a floorplan of a domestic home with the three cylinders distributed at different points throughout the home, suggesting there are 'scales to fit homes of any shape or size' and with an accompanying 'simple [smartphone] app to control what matters' (2016). In this video, the white cylindrical points are set against a stark white background, with white being a key marker of a desired modernist aesthetic (Wigley, 1995); the ad's layout and messaging work to convey a clean and sharp design and extreme simplicity of installation, promoting a vision of wireless connectivity that is a far cry from the magazines of the late 2000s that sought to demystify and explain to consumers the intricacies of wireless networking installation in lengthy articles and diagrammatic representations.

To return to Silverstone's account of domestication, he suggests that the next phase in the domestication process, appropriation, is a crucial stage vis-à-vis imagination. 'The act of purchase', he writes, becomes, 'potentially, a transformative activity, marking a boundary between fantasy and reality, opening up a space (or not) for imaginative and practical work ... on the meaning of the object, either as compensation for disappointed desire [where, for instance, Google Wifi fails to deliver on its promise of seamless connectivity] or as a celebration of its fulfilment' (Silverstone, 1994, p. 126). It is this second stage of domestication that we consider next.

Appropriation

Appropriation refers to the processes by which material objects come to enter a household. This process is initiated when households engage with the broader structures of signification that are associated with the consumption of consumer goods, whether this be by consuming advertising and marketing messages about material objects, or by researching particular goods or services. Appropriation proper occurs once households mobilize and act on their commodity knowledge (Lally, 2002, p. 50), and at the point at which an object 'leaves the world of the commodity ... and is taken possession of by an individual or household and *owned*' (Silverstone, Hirsch, and Morley, 1994, p. 21).

In what forms an important qualifier to the above understanding, Silverstone, Hirsch, and Morley go on to add that: 'appropriation is not confined only to material objects but crucially also applies to the appropriation of media content, to the selection of programmes to watch, computer software to buy, telecom services to subscribe to; though "ownership" of these things is of a different order from the ownership of objects' (Silverstone, Hirsch, and Morley, 1994, p. 22).

Even according to this expanded understanding, the process of appropriation struggles to accommodate Wi-Fi. This is because Wi-Fi, somewhat uniquely, performs a double articulation of the appropriation stage. First, households act on their commodity knowledge by appropriating (purchasing) a Wi-Fi router, such as a Linksys E3000. However, even this first form of appropriation is complicated by the fact that, increasingly, Wi-Fi does not necessarily enter the home in the way that, say, a TV, computer, mobile phone, or internet-connected fridge might. While the earlier discussion of magazine and catalogue advertisements suggests that Wi-Fi routers are purchased outright, this is often, in an Australian context at least, no longer the case, with the router tending to be bundled (along with USB cabling and other peripherals) as part of telecom and broadband data services. Where the Wi-Fi router is bundled with other services, we might say that Wi-Fi is smuggled into the home – it enters covertly and, thus, in this particular scenario, sidesteps the appropriation phase of the domestication process. Second, Wi-Fi's principal function is as a 'gateway' technology that provides access to a proliferating array of other devices. As David Morley (2017, p. 113) puts it, 'nowadays, the domestic living space is no longer simply "doubled" by the presence of the television screen but refracted in ever more complex ways by the simultaneous presences' of multiple screens and other internet-enabled devices, all of which are connected via Wi-Fi. As we noted in the earlier discussion of the imagination phase, such interconnectivity via Wi-Fi was being promoted very early on in Wi-Fi's history as a consumer product and service. The result of this double articulation is that Wi-Fi rapidly recedes from view as a good or service that warrants domestication in its own right, while multiple new connected devices enter the home that are in need of domesticating. In this way, the domestication process, when applied to Wi-Fi, might be more productively conceived of as a phased or multi-layered phenomenon, rather than as a single process in which each new household technology is domesticated in turn.

Having said this, once within the home, the Wi-Fi router – the enabler of Wi-Fi access – can and does become the focus of significant negotiation around its placement within the home (objectification), while Wi-Fi, especially access and data, forms the focus of intense negotiation concerning its position within the moral economy of the household (incorporation and conversion).

Objectification

Objectification references the process of physically and symbolically placing objects within the home. Where a technology is located within the home is a crucial element of the moral economy of the household (Silverstone, Hirsch, and Morley, 1994, p. 25), and is subject to much negotiation. This is the case whether that technology be a television (Spigel 1992; Morley, 1986), a personal computer (Lally, 2002), or a wired, high-speed broadband access point (Arnold et al., 2016). Within domestication research, much is made of how this issue of placement forms a complicated, delicate, and ongoing process. These negotiations tend to be fraught because the display of any household item does not occur in isolation, but forms part of an ensemble of objects that together convey 'an expression of the systematic quality of a domestic aesthetic' (Silverstone, Hirsch, and Morley, 1994, p. 23). Thus, any object – whether old or new – always appears, and is displayed, 'in an already constructed (and always reconstructable) meaningful spatial environment' (p. 23).

As with appropriation, objectification, crucially, is a process that applies not just to physical objects that enter the home, but also to 'semi-material artefacts', such as 'computer software, videos or the stuff of telephone conversations' (p. 24). While these things don't go on physical display, their objectification occurs through 'their incorporation into the temporal structure or fabric of the household' (p. 24).

How, then, does objectification function in relation to Wi-Fi? Harmeet Sawhney (2005) makes the point, already noted in the 'Imagination' section, that part of the early promise of Wi-Fi technology related to its potential for opening up opportunities for internal mobility within a building. He writes: 'Just as a cordless [handset] user can walk around the house ... while talking over the telephone, the early Wi-Fi networks sought to create similar mobility for Internet access via laptop computers' (p. 53). This continues to be a key part of the appeal, and now very much a taken-for-granted aspect, of domestic Wi-Fi use. As one article headline puts it, 'home is where the hand-held is' (Wagstaff, 2004). However, for such internal mobility to be optimized, where a Wi-Fi router is placed within the home is a prime consideration.

What emerges with respect to domestic Wi-Fi provision, then, is that objectification remains a key concern, yet this rarely takes the form of aesthetic considerations around display. Rather, key drivers behind how and where to accommodate Wi-Fi routers tend to involve more practical concerns, including around signal reach, so as to maximize internal mobility without compromising on signal strength. Such considerations are important given that modern domestic homes are often less than ideal environments for supporting optimal Wi-Fi network performance. Homes can contain numerous connectivity problem areas, or dead zones, as a result of stairs, corners, interference from other electrical appliances, or as a result of signal-impeding building materials (such as brick or thick timber walls).

A modem's firmware or hardware can also dramatically impact network performance. In a 2019 study of modem performance by the Australian Communications and Media Authority, it was found that '30% of modems could not hit 100Mbps on 2.4 GHz at a range of 5 metres', and that, 'once walls were put in the way, 26% of devices could hit 10Mbps on 2.4GHz, while 40% could achieve a data rate over 80Mbps using 5GHz' (Duckett, 2019).

Sub-standard household Wi-Fi performance and connectivity issues – or what Bly et al. (2006) refer to as the management of 'broken expectations' within the digital home, wherein there is a mismatch between what a person expects to be able to do and specific device capabilities and reliability issues – have given rise to an important secondary market for network performance-enhancing products, including boosters, extenders, mesh routers, and Powerline adapters. *Boosters* are products that come with more powerful antennas than a standard router; they are designed to be plugged into the wireless router via an Ethernet port, and to broadcast the incoming connection as an additional wireless signal. *Extenders*, which need to be placed part-way between the router and the blackspot area, plug into electrical outlets and, via antennas, receive, replicate, and extend a connection through a given space. *Mesh routers* use multiple devices to 'sling a speedy Wi-Fi signal to all corners of the house' (Crist, 2019). Meanwhile *Powerline adapters* use a combination of networking hardware and electrical wires to create a more reliable connection between an internet router and an end device, but with the disadvantage of making wireless devices wired again. Felicitous placement of Wi-Fi access points and of network amplification technologies is thus an important step towards domestication of Wi-Fi.

It is through their strategic placement that these network technologies also become key enablers of the proliferation of distributed and mobile media within the home. With this come changes in household patterns of habitation. The changing forms and patterns of occupation within households involving technology use facilitated by Wi-Fi (Nansen et al., 2011, p. 702) not only lead to an erasure of 'the lines distinguishing spaces within the house' (Wilken, Arnold, and Nansen, 2011, p. 6), but also, and crucially, to changes in the moral economy of the household. Reflecting on the shifts that had been observed in a single household they had visited on numerous occasions over a four-year period

(2004–7), Nansen et al. observe that, during early household visits, audiovisual entertainment was no longer consumed solely through the television, nor was viewing limited to the 'television room' (Nansen et al., 2011, p. 703). Rather, what they found was that 'viewing devices had multiplied: digital content could be consumed on different devices, in different places and at different times – previously set aside for activities such as eating or chatting – breaking down spatial, temporal, *and* sociotechnical distinctions' (p. 703). From this passage it becomes clear that provision of wireless networking has 'led to different modes of dwelling' (p. 704), with 'implications for domestic performances, routines, and practices' (p. 705).

What Nansen et al. (2011) also found was that, in addition to erasing the lines distinguishing spaces within the home, wireless networking was leading, increasingly, to an actual reconfiguration of the home itself – something that became evident from their study of households who were carrying out renovations or who had moved into designed homes. The picture that emerged was one where 'dwelling with wireless, distributed and mobile media practices made dedicated [media] spaces redundant' (p. 711). The preference was for 'a house structure that enabled [wireless] integration rather than one that enabled [spatial/media] differentiation' (p. 709).

This is an architectural lesson that the Wi-Fi Alliance also seems to have embraced enthusiastically. Describing home Wi-Fi as an expected utility, rather than an amenity, the Wi-Fi Alliance has developed a certification programme, called Wi-Fi Home Design, for new home builders – such as the Florida-based construction company Lennar Corporation (Lennar, 2019) – who wish 'to offer built-in Wi-Fi networks with comprehensive coverage throughout the home, as well as outdoor living spaces' (Wi-Fi Alliance, 2019a): 'Akin to builder-installed lighting schemes, turnkey Wi-Fi Home Design networks are professionally designed and implemented, bringing reliable, scalable, high-performance Wi-Fi to all parts of the home' (2019a).

One difficulty with the implementation of such schemes, as Matthew Fuller (2007, p. 1) notes, is that 'complex objects such as media systems' should only ever be understood as involving 'processes embodied as objects, as elements in a composition' that settle 'temporarily into what passes as a stable state' before reforming and resettling, and so on, in a process that is ongoing. The fact that this home certification system is based on Wi-Fi 5 (Wi-Fi standard 802.11ac), just as Wi-Fi 6 (802.11ax) is being heralded (Hoffman, 2019), provides a case in point. It is in this respect that Google, with its movable 'points' for flexible home wireless networking, may have stolen the march on the Wi-Fi Alliance.

Incorporation

Whereas objectification is focused on spatial aspects, incorporation emphasizes the temporalities associated with the moral economy of the household (Silverstone, Hirsch, and Morley, 1994, p. 24). That is to say that, while objectification is principally about *where* a technology is located, incorporation is about *how* it is to be used and *by whom*, and how it fits within the routines of domestic family life (p. 25).

These issues of access and control are played out in a number of different ways with respect to Wi-Fi. One key way in which they surface is around the bounding of agency of children within the home (Haddon, 2016, p. 25), particularly in relation to ICT use. In an Australian study of children's use of mobile devices, for instance, Bjørn Nansen found that parents used a variety of measures to control screen access. These ranged from 'traditional mediation processes involving the imposition of rules around media free spaces or times, through to the use of reward-based limits for good behavior, as well as more technology-focused ways of restricting access such as hiding devices, changing passwords, or turning the

Wi-Fi off' (Nansen, 2019, p. 38). Wi-Fi access also figured in a US study examining a similar set of concerns. In Rebekah Willett's study of children and gaming, she found that, among her participants, 'children with tablets did not have 3G or 4G services' (Willett, 2017, p. 156). The result was that 'they could not access the Internet "on the go", so they only accessed the Internet from home or occasionally in relatives' homes if they had set up the WiFi connection' (p. 156). While far less common, Wi-Fi router reconfigurability also presents a more dramatic method, for the tech-savvy parent, of 'extensively modify[ing] medium access control' (Mackenzie, 2010, p. 113).

In taking such measures as these, parents are not just trying to assert control over device use. Rather, these negotiations around Wi-Fi access and use are closely intertwined with parental desires to influence and exert control over the moral economy of the household. As Silverstone, Hirsch, and Morley put it:

> The politics of the family and the neighbourhood, the conflicts over ownership and control of (*inter alia*) information and communication technologies and the status of the family or household members are all expressions of, as well as elements within, the continuous work of social reproduction that provides the basis for the coherence of the household's moral economy. (Silverstone, Hirsch, and Morley, 1994, p. 25)

It is in the above context that we might also begin to make sense of disputes between neighbours that are played out via Wi-Fi network address names, including over broadband access ('Go Away Don't Steal My Broadband', 'Covet not thy neighbour's wi-fi') (Heyden, 2012). While these seemingly petty conflicts often appear humorous to those not involved, they nonetheless form a further illustration of how Wi-Fi access and control are entwined with the ways that 'spatial and temporal boundaries are created and defended within and around the household' (Silverstone, Hirsch, and Morley,

1994, p. 25), especially through our use of information and communication technologies.

Conversion

The last of the five elements of the domestication framework is conversion, which pertains to the process of defining the relationship between the home and the outside world via the technologies that are consumed within the household. This is the most abstract of the five elements, and develops the idea that, once commodities are brought into the home, they must undergo a form of conversion and transformation if their display, use, and significance within the home are to have wider purchase and meaning outside of the home. Silverstone, Hirsch, and Morley (1994, pp. 24–5) illustrate this point via the historical example of television: 'Television is the source of much of the talk and gossip of everyday life ... The content of its programmes, the twists of narrative, the morality of characters, the anxieties around news, provide in many places and for most of us, with greater or lesser degrees of intensity, much of the currency of everyday discourse.'

At the heart of the idea of conversion, and as conveyed in the preceding quote, is the suggestion that communication technologies perform a 'double articulation', in which they are 'facilitating conversion (and conversation) as well as being the objects of conversion (and conversation)'. As with appropriation, Wi-Fi does not appear to map neatly onto this concept of a double articulation. As a gateway technology that connects other devices to each other and to the internet, Wi-Fi is rarely an *object* of conversion and conversation (unless it fails), whereas it does play a crucial role in *facilitating* conversion and conversation. Lynn Spigel, writing in 2005, just as domestic Wi-Fi was beginning to burgeon, notes that, 'with mobile phones, personal digital assistants (PDAs), laptop computers and the like we increasingly experience being at home while in public and we also experience being in public while at home' (Spigel, 2005, p. 414; see also Morley, 2003). This experience of being in public while at home, and having the resources to do so, is something that is felt particularly keenly by teenagers. In an Icelandic study of home internet use among those aged 12–15 in the greater Reykjavik area, Jodie Birdman (2016) found that, among this cohort, agency, belonging, and empowerment were all, indirectly yet firmly, tied to having a 'really good internet connection' (p. 75). As one participant summarized it, if you don't have Wi-Fi (and snacks) at your house, you don't have friends – 'that's it basically' (p. 75). For these teens, not having home Wi-Fi is not only regarded as unusual or remarkable – a kind of reverse or negative conversion phase – it is considered a form of social suicide. This is due to the fact that, as Silverstone (1994, p. 130) reminds us, conversion 'is an indication of membership and competence in public culture, to whose construction it actively contributes'. Access to Wi-Fi, for younger teenage householders, can form a key marker of such public competence and social acceptability.

Much of the classic domestication framework scholarship and examples tends to focus on certain sorts of devices. These tend to be stand-alone and often big-ticket items, such as the television or desktop computer (Berker et al., 2006, p. 14). However, a great deal has changed since then, and the household has become a site of media and technological proliferation. As noted earlier in this chapter, 'nowadays, the domestic living space is no longer simply "doubled" by the presence of the television screen but refracted in ever more complex ways by the simultaneous presences' (Morley, 2017, p. 113) of multiplying screens, multiplying television-related hardware – including set-top boxes and digital recorders (Hesmondhalgh and Lobato, 2019; Meese, et al. 2015), games consoles, and the like – and innumerable other internet-enabled and wirelessly connected devices. Our argument to this point in the chapter has been that the arrival of Wi-Fi within the home has facilitated and hastened many of these transformations and has also complicated critical understandings of the domestication process. In the section that follows, we discuss more explicitly the rise and development of the contemporary smart or connected home. The argument we develop is that the enabling and sustaining capacities of Wi-Fi become even more vital for the successful operation of the technologized, connected home. Yet, clearly, Wi-Fi as a technology has had to undergo certain adjustments and advancements in order to facilitate and keep up with proliferating digital devices and connected things.

The connected home

The phrase 'smart home' is principally used to describe an integrated system of 'technological convergence' within the architectural setting of the domestic home. Technological convergence in this instance refers to the interconnection of various household appliances, telecommunications equipment, environmental control devices, and security systems, and so on, all of which are fed into and controlled via a central processing unit and connected to the internet. The common vision of this technology, at least as it has historically been presented in the popular press, is of a 'smart' or 'intelligent environment' that not only controls various environmental factors (such as lighting and heating) and responds to security breaches, but also pre-empts user needs (such as in Nicholas Negroponte's vision of your refrigerator notifying your car to remind you that you are out of milk – Negroponte, 1995, p. 213).

As Lynn Spigel (2005, pp. 405ff.) reminds us, the contemporary 'smart home' idea is merely the latest version within a much longer history of 'homes of tomorrow' (see also Spigel, 2010; Aldrich, 2003). And when considered as part of this longer narrative, there are numerous historical precursors to the 'smart home' idea – including Charles Barry's 1839 London Reform Club with its 'largely invisible heating, ventilation and systems of mechanical communication buried within the walls' (Forty, 2004, p. 89), William Randolf Hearst's opulent 1925 California castle San Simeon with its multiple radios tuned to different frequencies and piped into his private suite (Gates, 1996, p. 248; Coffman, 2003), as well as US art collective Ant Farm's 1973 experimental House of the Century project in Texas (Lewallen and Seid, 2004), to name just a few.

The term 'smart home' usually refers to 'a residence equipped with computing and information technology which anticipates and responds to the needs of the occupants, working to promote their comfort, convenience, security and entertainment through the management of technology within the home and connections to the world beyond' (Aldrich, 2003, p. 17). Arguably the paradigmatic modern early example of the 'smart home' is Bill Gates's much-publicized US$35-million residential complex near Seattle, Washington. Of all the domestic technologies this house is said to contain, the one Gates spends most time detailing in his own written account of the project is the electronic pins visitors to the house are given upon entering. 'The electronic pin you wear', he writes, 'will tell the house who and where you are, and the house will use this information to try to meet and even anticipate your needs – all as unobtrusively as possible' (Gates, 1996, p. 251). This may involve re-adjusting environmental settings to visitors' preferred temperature and light levels, as well as catering to individual audiovisual tastes of each guest as they move from room to room. Beyond the Gates house example, contemporary manifestations of the 'smart home' include, along a scale of complexity, high-end bespoke systems requiring substantial investments of time, money, and resources to implement them at one end (Waddell, 2018), and homes that have one-or-more off-the-shelf items – such as a Nest thermostat for instance – that sit alongside a plethora of other connected devices (TV, smart-phones, etc.) at the other end.

Richard Harper, in his 2011 book *The Connected Home*, argues that the smart home idea has, to date at least, largely failed to live up to its hype around the promised provision of anticipatory automation and control. (The smart home vision, as others have argued, has also been built around decidedly problematic visions of labor and their gendered logics – see Strengers and Kennedy, 2020; Fortunati, 2018; Spigel, 2005.) The reality, Harper argues, is far messier and more mundane than smart home visions generally permit. It is a reality centred around already existing and ongoing individual and social uses of an ensemble of technologies and services made possible by various forms of connective technologies that are now an integral part of the home environment and that have had a significant effect on the home. Harper writes: 'The technologies that make these connections are remarkable in their range and in their consequences; they are remarkable too for being commonplace in living rooms, kitchens and bedrooms. They include ... online games technologies ... ; video connections, via Skype ... ; and even technologies that don't seem about connection, but which rely on connection to function adequately, such as e-readers' (Harper, 2011, p. 8).

Even more remarkable still is that a single technology – Wi-Fi – sits at the heart of the connected home and enables the very forms of connectivity listed above by Harper. Wi-Fi is fundamentally a connectivity-enabling technology (Webb, 2011). It is through the domestic gateway of the Wi-Fi router that data enters and exits the home; it is through this same gateway that the domestic internet of things also comes into being. And the support Wi-Fi gives to the domestic internet of things is a mutually reinforcing process: as acceptance grows for it as the standard household protocol for domestic technology connectivity, as common electrical and electric home appliances – such as televisions, fridges, thermostats, coffee makers, and light bulbs – have in-built Wi-Fi capabilities added, the more it is that home Wi-Fi accelerates the introduction and proliferation of 'smart home devices'

(Madasu, 2016; Weinreich, 2017). One estimate puts current ownership of Wi-Fi-enabled devices at nine per household, on average, with predictions this will rise within several years to an average of fifty (Kastrenakes, 2019). And industry predictions are that the global market for smart home devices is expected to grow 26.9 per cent year over year in 2019, to 832.7 million shipments (IDC, 2019).

Over the past fifteen years, then, Wi-Fi has developed so that it now takes on an integral role in facilitating technological connectivity within the domestic home. Yet Wi-Fi is by no means a static technology. Through the stewardship of the Wi-Fi Alliance, Wi-Fi is continually evolving to meet the demands of the complicated role it plays within the domestic home. For instance, there have been several developments in Wi-Fi standards targeted at increasing both the range and speed of domestic Wi-Fi coverage, and enabling high network use while avoiding network congestion within the home. Here we briefly survey a number of these – HaLow, AF, AD, and Wi-Fi 6.

In 2016, the Wi-Fi Alliance announced the 802.11ah (HaLow) standard (which now forms part of the rebadged Wi-Fi 5), specifically to cater for IoT sensors that don't require high data rates. HaLow operates on the 900 MHz band, and is meant for wider-range data transmission, such as connecting things like thermostat sensors at opposite ends of a house. AF, a related standard that is otherwise known as 802.11af, employs unused television spectrum frequencies – known as 'white spaces' – in UHF and VHF to transmit information, in frequencies that fall between 54 MHz and 790 MHz. Like HaLow, AF can also be used for low-power, wider-range applications. Then there is AD, the 802.11ad standard, which operates on the 60 GHz band. This is considered ideal for very high data rate but very short-range communications. The AD standard is pitched at those wishing to transmit high-definition and 4K video wirelessly, using the 60 GHz band through products such as WirelessHD, WHDI, and WiGig

(LinkLabs, 2018). While the 60 GHz band is less congested, which means that it can transfer more data at once, the shorter wavelengths equal a shorter range (although this can be extended up to 10m using beamforming technology) and experience difficulty travelling through walls; it is also a standard that has been slow to receive widespread commercial support (Hoffman, 2018; Morrison, 2016). These are just a few examples of proliferating 802.11a class standards that have been developed over time to address the changing needs of domestic broadband users.

A major step change in domestic Wi-Fi provision followed the development of Wi-Fi 6, which was ratified in late 2019. The Wi-Fi 6 standard is said to lower latency, increase data transfer speed, and reduce likely domestic Wi-Fi dead zones thanks to beamforming (whereby the router focus signals on a particular device to improve the strength of a connection), and to have increased security (through the use of the WPA3 security protocol). Wi-Fi 6 is also said to improve device battery life through the use of Target Wake Time (TWT). TWT works by permitting devices to schedule communications with the Wi-Fi router, reducing the need for these devices to keep their antennas powered on to transmit and search for signals (Kastrenakes, 2019).

A key focus of Wi-Fi 6 is on improving network performance when multiple devices are connected simultaneously, and on handling competing requests for network access (Marr, 2020). It achieves this through the integration of a range of new technologies and standards, including MU-MIMO functionality, OFDMA, BSS 'colours', and RTT. The first of these, multi-user multiple-in/multiple out (MU-MIMO) functionality, allows a router to communicate with up to eight devices at the same time (Kastrenakes, 2019). The second, orthogonal frequency division multiple access (OFDMA), enables a single transmission to deliver data intended for multiple devices at once (Kastrenakes, 2019). The third technology permits wireless access points near each other to be configured to have different basic service set (BSS) colours or channels, numbered 0 to 7. The process works as follows: 'If a device is checking whether the channel is all clear and listens in, it may notice a transmission with a weak signal and a different "colour". It can then ignore this signal and transmit anyway without waiting, so this will improve performance in congested areas, and is also called "spatial frequency re-use"' (Hoffman, 2019).

The last of the aforementioned technologies, Wi-Fi round-trip time (RTT), refers to 'the length of time it takes for a signal to be sent plus the length of time it takes for an acknowledgment of that signal to be received' (NetSpot, 2019). What is important about RTT is that it enables smartphones and other devices to use time-of-flight instead of signal strength to figure out how far away they are from Wi-Fi routers. Thus, with a single Wi-Fi router, the distance to the device is determined. With three or more Wi-Fi routers, trilateration is achieved, with an accuracy of 1–2 metres (2019). Wi-Fi RTT has been described as a 'game-changer' when it comes to home automation (2019). All of these technologies combine within Wi-Fi 6 to improve overall network performance, and, particularly, to cater for the growing array of networked devices populating the contemporary connected home.

The final emergent technology that warrants mention here is 5G (fifth-generation) wireless cellular technology. Unlike 3G and 4G, 5G is not a standard upgrade to infrastructure and mobile networks. While it is said to deliver much faster data transmission due to increased bandwidth and improved antennas, it will also lift network capacity, such that an array of systems and services, from utility grids and industrial machinery to cities, cars, and homes, can all be connected through wide-area networks. This will enable telecommunications providers to expand beyond the provision of voice and online communications to become the technological basis for automated technologies that support smart homes. The technology also allows some providers (such as Optus in Australia) to offer high-performance alternatives to fixed wireless and wireline services. Despite all of this, and the significant wider impacts and innovations that the arrival of 5G will herald, it is unlikely that 5G will replace Wi-Fi any time soon as the domestic 'gateway' technology of choice. This is due to Wi-Fi's present certainty and stability, the significant existing investments telecommunications companies have made in fixed-line services, the cost-effectiveness of installing Wi-Fi within the home, the high costs associated with building extensive 5G blanket infrastructure and installing femtocells within domestic homes so that 5G services can operate efficiently and at high transmission rates indoors, and the fact that so many consumer electronics goods and connected devices already come Wi-Fi-enabled (Apostolopoulos; 2019; Richards, 2018; Santo, 2017; Wall, 2020). What is more likely is that 5G and Wi-Fi will, for the most part, co-exist as complementary rather than competing technologies, with mobile devices handing over to 5G networks once outside the home.

To return to processes of domestication, it is worth noting that, while Wi-Fi 6 (alongside other Wi-Fi standards) is designed to accommodate a burgeoning array of Wi-Fi-enabled household technologies, bringing new routers and new connected devices into the home adds further layers of complexity to the domestication process as a whole. Just as Wi-Fi must be 'domesticated', so too must all of these IoT 'smart' technologies. Each carries its own objectification, incorporation, and conversion considerations. Reconfirming what is already established from the domestication literature, Tom Hargreaves and Charlie Wilson (2017, p. 88) note that, 'on top of requiring domestication in and of themselves, smart home technologies also demand that many other aspects of the domestic environment are re-domesticated into the new "smarter" home'. Furthermore, householders have been found to adopt a range of adaptation strategies. These strategies encompass non-use, resistance, rejection, as well as partial use, so as to render IoT technologies more familiar and less disruptive – a process that Hargreaves and Wilson (p. 88) refer to as '"shallow" domestication'. Major contributors to 'shallow domestication', they argue, are the labour involved in setting up and maintaining connected technology systems, and the lack of support available to householders for carrying out the considerable work involved in domesticating connected devices and networked systems (p. 88). The effort involved in establishing and sustaining Wi-Fi-enabled connected home systems is the inconvenient truth of automated home systems – indeed, as Richard Harper (2011, p. 4) notes of Gates's Seattle home, 'while it would appear to offer smart home-like functions, this is achieved by having a full time support staff: the smartness here is delivered through Wizard of Oz techniques'. Despite improvements in Wi-Fi standards, proliferating connected devices, it would seem, generate proliferating domestication challenges (see Kennedy et al., 2020).

Conclusion – Wi-Fi and the domestication and transformation of the home

In this chapter, we set out to investigate whether, and how well, home Wi-Fi fits with, and is comprehensible through, the domestication framework as conventionally understood. What is clear from this examination is that, in certain respects, the home Wi-Fi router, especially when regarded as a stand-alone technology, can be understood as subject to some of the same phases of domestication (e.g. imagination, incorporation) as other items, such as TVs and desktop computers.

In other respects, though, home Wi-Fi pushes the limits of the domestication framework, and exists on the edge of legibility of domestication theory. This occurs in a number of ways. For instance, Wi-Fi routers often enter the home bundled with data services, suggesting there is not always a clear appropriation phase. Within the objectification phase, there is also often a tension between the home Wi-Fi router being a clearly visible technology (where it is often positioned less for purposes of display than for best signal coverage and signal strength) and it being hidden away from view (as an unsightly box that detracts from the aesthetics of the home environment). In addition, Wi-Fi doesn't always map neatly onto the conversion phase. While Wi-Fi rarely forms an object of conversion and conversation (unless it fails), not having Wi-Fi – which might be cast as a kind of inverse or negative conversion – is regarded by some younger householders as socially disastrous.

Wi-Fi also exists on the edge of legibility of domestication due to the way that it serves as both a domestic information and communication technology *and* as an enabling infrastructure. Over recent decades, multiplying media and information and communication technologies have come to form a vital part of the fabric of the contemporary household. In Silverstone, Hirsch, and Morley's (1994, p. 15) words, 'they provide, actively, interactively or passively, links between households, and individual members of households, with the world beyond their front door, and they do this (or fail to do this) in complex and contradictory ways'. Wi-Fi has come to form a crucial facilitator of, and mediator in, the creation and sustenance of these links.

In this chapter, we have suggested that the notion of the gateway provides a productive means of grasping how Wi-Fi fulfils this mediating – and infrastructure-enabling – role. Just as a front gateway provides pedestrian (or vehicular) access to, or exclusion from, a domestic home, Wi-Fi can be seen to fulfil related functions. As noted earlier, it is literally through the 'residential gateway' (the Wi-Fi router) that network access into and out of the home, and between connected devices within the home, is managed. It is as a gateway technology that Wi-Fi in fact plays a crucial role in *domesticating* the home. Wi-Fi fulfils this domestication role quite specifically in that it creates a digital space for household internet. The Wi-Fi router, as residential gateway, both serves as point of entry into a bounded space and creates a boundary for wireless signals. The residential gateway is a manifestation of a boundary, with people inside and people outside, and it is through this that control is exercised. Wi-Fi creates a digital territory in and around the home. It does this literally, by defining a communications zone, and maps it (im)perfectly onto domestic space. This is why access to, and control over, the residential gateway of the home Wi-Fi router can be so hotly contested. It is via this device that the communications environment of the home is delimited; it is also via this device that data access within the contemporary home is negotiated between households, among individual members of households, and with the world beyond their front door. Wi-Fi, in short, forms a literal and figurative gateway around and through which the wireless boundaries of the home and the moral economy of the modern, connected home are defined, negotiated, and defended.

In addition to domesticating the home, Wi-Fi has also played a crucial role in modern times in transforming the home. What we are pointing to here is a recent acceleration of a longer historical process of transformation concerning the things that we make, consume, and share within the home. The kinds of transformations we are alluding to here can be captured in the following two examples. The first is the transition over time from the collecting and keeping of family recipes, which Leong (2013) describes as repositories of household knowledge, to the gradual emergence of commercially produced cookbooks that are purchased and consumed by householders as supplements or replacements to family-generated recipe books (Notaker, 2017), and, more recently still, online food-related platforms (Lewis, 2020). The second is the transition over time from the taking of analogue photographs and the filing of them in family photo albums, which Sandbye (2014) regards as crucial artefacts of domestic material culture that serve vital communicative and interpretative functions within the life of the household, to the supplanting of some of these functions by the taking and sharing of digital images online (Palmer, 2010). These shifts are reflective of a changing relationship between household audio-visual economies and wider market audio-visual economies, and of 'the global circuits of exchange that now constitute our homes' (Johnson and Lloyd, 2004, p. 157). This is to say that the entrance of technology into the home opens up new forms of interaction, wherein informal processes and practices have evolved to become more managed and controlled processes that are handled through digital platforms. The arrival of Wi-Fi within the home has thus not just led to the proliferation of contemporary connected devices within the home, but also contributed to the acceleration of these longer, historical transformations within the economic space of the home, which, once quite informal, is now becoming very formalized.

4

Community

During the early 2000s, Wi-Fi left the confines of buildings and became available to users in outdoor spaces. Groups of committed geeks climbed rooftops and adapted antennas in efforts to concentrate signals and beam them over greater distances. In the process they created the first community Wi-Fi networks, relying on the generosity of individuals and businesses who donated unused bandwidth, equipment, access to physical structures, and some empty food containers (for 'can-tennas'). Through community coordination and technical ingenuity, Wi-Fi became publicly available across multiple city locations for the first time.

Community Wi-Fi networks are created by people with a shared interest that they themselves have defined, as opposed to a contractual or rules-based relationship with a government or commercial provider (Gurstein, 2007). However, the social coordination that resulted in these Wi-Fi networks is similar to that associated with open source software, defined as a commitment to openness, sharing and building on the work of others. In the case of Wi-Fi, this ethic may also be carried through into hardware design and deployment, and reflected in the rules placed on the network, such as terms of service and authentication software. These networks also hold other properties associated with the concept of community, including distinctly local characteristics, which may produce shared concerns such as the maintenance of resources in a defined geographic area.

In Chapter 2, we discussed how it came to be that users can make decisions about where and how Wi-Fi is used. In this chapter, we discuss community Wi-Fi networks, including their rise as a phenomenon in the early 2000s. Some communities, upon learning that Wi-Fi is more adaptable than other networking technologies, have exerted control and preferences for internet provision using Wi-Fi. In the final part of the chapter, we look at how community dynamics can play out even where the network is provided by government or commercial operators.

Community

The word 'community' has multiple meanings. It can refer to a group of people who reside in geographical proximity to each other, typically with some shared interests in local issues, facilities, and systems. A community can also be a group that has distinct characteristics, such as a diasporic community that has a shared language, culture, and history. Or it can mean a group that comes together around a particular interest or skill, with the intention to develop that interest or skill through mutual practice and learning, known as a 'community of practice' (Lave and Wenger, 1991).

The term 'community' is sometimes associated with a sense of togetherness, emphasizing things in common and assuming harmony and social cohesiveness. However, communities can also be transient, fractured, and aligned over causes and beliefs that are harmful to others. A community can be defined by its borders – by those it excludes, rather than who or what it embraces (Young, 1986).

Community *organizations* are groups of people who mobilize, cooperate, and work together for reasons other than financial incentives. They may do so in unstructured ways or develop formal rules that determine how decisions are made. When communities formalize into not-for-profit associations, they are required to comply with certain laws and reporting requirements. Such structures thereby bring the sometimes unwieldy and amorphous field of relationships and group identities under the purview of the state (Rose, 1999). When organized around corporate structures, communities can be as large and effective as private companies, yet they differ from private companies as profits are not distributed to shareholders but directed back into the organization. Taken collectively, not-for-profit enterprises are sometimes referred to as the 'third sector'.

There are various reasons why people would cooperate or give their time to tasks that they might not personally profit from. In the case of Wi-Fi, the motivations underpinning community efforts are diverse – ranging from technical curiosity and learning through a community of practice, to assisting others to experience the opportunities of the internet in areas where it is not available. In the field of communication technology, this form of social coordination has produced outcomes that have benefitted many, including people who would otherwise have missed out on internet services at the time.

Community infrastructure cooperatives

Community organizations, sometimes referred to as cooperatives, often arise when commercial and public services are absent or meeting only the needs of a minority. Two examples of large-scale communication networks – one historical and one contemporary – illustrate how this can come to pass.

> *Example 1: It is the United States in the 1920s. Automobiles are beginning to outnumber horses with the rise of Henry Ford's Model T, and new communication technologies – film and radio – are creating a mood of novelty and newness in the cities. However, if you are part of a farming household, you are more likely than an urban household to have a telephone; 39 per cent of households in rural areas, compared to 33 per cent or less in urban areas (Fischer, 1992), have a working telephone. In the previous decades, telephone companies have chosen to focus only*

on people living in cities, believing that people living in rural areas are fearful of technology. By the 1920s, the rural households whom the private telephone companies had dismissed have come up with their own solutions.

In his history of the early telephony in rural America, Claude S. Fischer (1992) tells of how, after being denied a service by a commercial company, farmers would create telephone systems between farms by wiring fences. Anywhere between fifteen and fifty farmers, as well as doctors and merchants, would then form a mutual stock company (a company that exists to provide services for members and does not seek to make profits). Others could subscribe to and receive a line and telephone. These local networks would arrange connection to either a larger mutual or a commercial network in town, enabling them to connect with the wider world. Women on farms would serve as switchboard operators during the daytime. In industry magazines, telecommunication companies bemoaned these networks, stating that they were inferior in their connecting lines, and lax with maintenance and finances. In addition, industry spokespeople complained that farmers used their networks in unacceptable ways, listening in on each other's calls, 'thereby weakening the signal and starting fights' (Fischer, 1987, p. 11), as well as gossiping and playing musical instruments on party lines. For Fischer, this particular history 'shows more human agency, more creation of supply by consumers, more popular modifications of a technology's use than would be suggested by conventional discussions of technological modernization' (p. 16).

> Example 2: Cuba, late 2018. The government telecommunications provider has just launched the country's first 3G network. Cubans have until now experienced very limited internet services. Despite the fact that internet adoption purportedly grew from 1.8 per cent in 2001 to 38.8 per cent in 2016, the time people spent on the internet was restricted by access, cost, and poor speeds. When the socialist Castro regime invested in a deep-sea cable in 1996,

> connecting the country to the public internet, it placed considerable limits on access and use. For instance, the 1996 Decree Law 209 confined internet access to approved locations – originally universities and selected workplaces, and later to state-run internet centres (from 2013) and Wifi hotspots (from 2015) (Kalathil and Boas, 2001). When the 3G network is launched in December of 2018, it is priced to be beyond the reach of many citizens, with a 4GB data package costing the equivalent of a monthly salary (Wyatt, 2018).
>
> However, some people living in Cuba are still able to participate in activities such as social networks and multiplayer online games at home without an internet connection. The SNET developed in the early 2000s as one workaround alongside a wider and tolerated informal economy, including international content shared by external hard drives (see Pertierra, 2012; Venegas, 2010). The gamer community built SNET originally as a means to run multi-player games such as Counter-Strike from different computers. The network consists of antennas and nodes that enable people to access content in much the same way as the public internet: searching, selecting (sending a request) and being routed to its host. A 2016 Cuban lifestyle magazine article stated that there were 8,000 computers, 'crisscrossing the capital from Cojímar to the east of Havana to the town of Bauta in neighboring Artemisa Province to the west' (Falls, 2016, p. 59). Developers have created SNET clones of Instagram, Facebook, and Reddit, which under SNET's terms of service must not contain advertising and be free to access. These strictly non-commercial rules mean that SNET remains legal and operational under Cuban law (Martínez, 2017).

The first example occurred during the emergence of industrial capitalism, while the second is happening in a planned economy under a communist government. What the early US telecommunications cooperatives and SNET have in common, however, is that both were provided to fill a need that was not being met by government or private firms. Those who participated in building the networks did so partly for their own benefit, but also for the benefit of others. The result, in both cases, was a bottom-up infrastructure designed according to what a group desired or required. The fact that these examples arose in vastly different political situations demonstrates the amorphous nature of community endeavour; community efforts span many political persuasions, or they can be apolitical. The early US telephone cooperatives were formed in response to the behaviour of the telecommunications industry, emerging or accelerating when patents expired and as a reaction to telecommunications executives disregarding the needs of rural households and businesses. Like these early telecommunications cooperatives, community W-Fi networks typically involve participants acquiring equipment and expertise and coordinating among themselves to provide services to a local area. Many community Wi-Fi initiatives, including SNET, exist where there is a perceived lack of suitable or affordable options.

The rise of community networks

In the early 2000s, skilled technologists in cities around the world formed groups and began to design and build cooperative Wi-Fi infrastructures. The stories of these networks follow a similar pattern. A small group of interested individuals would develop methods for purchasing equipment, installing authentication software, and experimenting with how far they could transmit signals. They would then produce information to assist others to join the network. As the networks developed, people also mapped the locations of members or participants (Wi-Fi mapping became a community of practice of its own). Face-to-face meetings were held to share and develop technology strategies and to introduce new members to the networks. Some pioneered new protocols and hardware that became industry standard.

For instance, NYCwireless developed at a time when the 'commercial battleground' for outdoor paid wireless hotspots in Manhattan was unfolding. NYCwireless set out to provide a free alternative in public parks by using guerrilla tactics (Forlano, 2008). One of the founders, Anthony Townsend, describes the group's first meetings in 2002, which: 'began in the early evening with demos and discussions about new wireless gadgets. They ended, as often as not, well past midnight over beers at a downtown bar. Around tables strewn with empty glasses and bottles, a dozen or more geeks would stay up late making plans to spread free networks throughout the city' (Townsend, 2014, p. 126).

Volunteers hung antennas from apartment windows and shopfronts – a practice that cost Townsend his job as a researcher for NYU after it was discovered that he had set up a wireless antenna out of the window of his university office on Washington Square (Townsend, 2014).

A key moment in the development of community Wi-Fi was when groups of technically minded people realized that Linksys, a manufacturer of networking technologies, had used Linux software that was licensed under the GNU General Public licence in its access points. The licence meant that others could use and adapt Linksys's firmware for different purposes (Song et al., 2018). For instance, in the early 2000s, a resident of Leiden in the Netherlands, Jasper Koolhaas, discovered that indoor Wi-Fi equipment could be adapted to build an outdoor free wireless network. Koolhaas teamed up with Linux user Marten Vijn and began adapting regular Wi-Fi devices – 'patching firmware, writing, and adapting device drivers for Linux' (Van Oost, Verhaegh, and Oudshoorn, 2009, p. 192) and experimenting with antennas to make them weatherproof and to extend their range. When they could not get distant nodes to connect, they brought in two amateur radio experts who amplified signals using bi-quad antennas which could bridge several kilometres in a line-of-sight situation. They then adapted the indoor-use modems by making them weatherproof, using drainpipes and plastic lunchboxes.

In her detailed ethnography of the ISF (Île Sans Fil, meaning Wi-Fi Island) in Montréal, Alison Powell documented it from its formation by a group of highly skilled geeks (25–35 people in 2005) to it becoming a formal not-for-profit company and public internet provider. According to Powell, ISF helped to define the technical, organizational, and symbolic practices of community Wi-Fi, and 'has influenced discussions on wireless applications for local communities in the national and international context' (Powell, 2012, p. 207).

In the case of ISF, the collective created their own Linux-based open source configuration and management system that also provided a landing page upon which local content was made available. As told by an ISF member, Michael Lenczner on the Linux blog, ISF developed Wifidog after a group in Seattle discovered a router could run Linux, and so 'the hacking began' (Lenczner, 2005, p. 3). At the time of Lenczner's article, ISF had 6,000 subscribers and was growing at a rate of 4–6 hotspots and 1,000 subscribers per month (2005). He writes that, 'From the beginning, ISF viewed setting up free hotspots as only a first step. The volunteers now had the tools to draw laptop users from their basements and home offices into public spaces. The next step of the project was to use the network of hotspots to help create a sense of local community' (p. 6). They did this by designing Wifidog to allow for location-specific content. Working with local new media arts groups, each Wi-Fi hotspot would be accessed through a landing page that had its own location-based arts, and the system ensured that a user would see new content each time they logged in. Wifidog was adopted by other Wi-Fi groups, including WirelessLondon, New York City Wireless and Paris Sans Fil. Powell writes that it was also taken up by commercial providers.

Similar Wi-Fi networks appeared around the world at the same time. According to Sandvig (2004), there were fifty-two Wi-Fi co-ops across the US, Canada, UK, Ireland, New Zealand, and Australia in 2003. In 2014, one estimate suggested there were over 400 community Wi-Fi groups (Jungnickel, 2014). Well-known networks included Seattle Wireless and the Athens Metropolitan Wide Network, which grew to more than 2,400 active nodes by 2011, as well as guifi. net in Spain (35,000 active nodes in 2019 – see https://guifi. net/en/node/38392) and Freifunk Germany (43,000 access points in late 2019 – see www.freifunk-karte.de). These networks evolved with different characteristics and objectives, reflecting local needs and interests. Jungnickel, in her study of community Wi-Fi, quotes one member of Australian Wi-Fi collective Air-Stream: 'We are building Ournet, not the internet' (2014, p. 4).

Warchalking

Not all grassroots activities are effective, and some can be problematic, as demonstrated by the short-lived warchalking phenomenon. Warchalking involved drawing symbols on buildings where households or organizations were accessible without password access (deliberately or by accident). The movement commenced in London in mid-2002, with symbols created and distributed by Matt Jones, based on US Depression-era 'hobo signs' that indicated where resources, work, and shelter could be found (Sandvig, 2004). The idea spread through mainstream media coverage, with the *New York Times* calling it one of the best ideas of 2002, causing its spread into many major cities. A related practice, wardriving, was undertaken by cruising the streets to detect insecure wireless LANs using homemade antennas. Jones reported to ZNet that a wardriving competition to find the best antenna had been won by a can of meat stew (Wearden, 2002).

Warchalking and wardriving can be interpreted as an expression of shared values around the open internet, or at least as a branding strategy for those who wanted to promote the idea. There was also an element of community in this coded system, expressed through affiliation via online forums. Although it only required entry-level hacking skills, the practice was not particularly reliable, in that chalk marks could easily be removed or washed off. Some pointed out that

Community 91

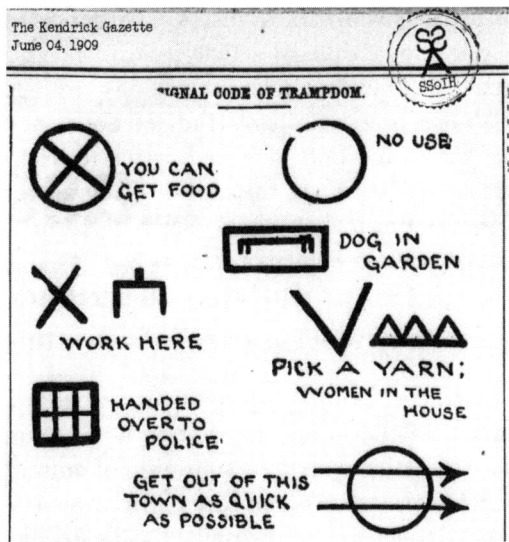

Figure 4.1 'Signal Code of Trampdom', in the *Kendrick Gazette* (Kendrick, Idaho), 4 June 1909.

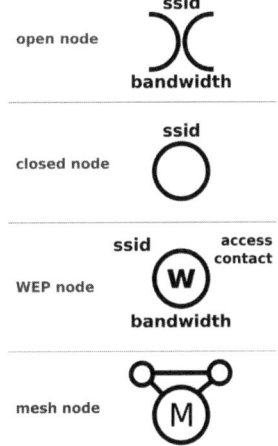

Figure 4.2 Warchalking symbols. Top to bottom: open node, closed node, WEP node, mesh node. The symbols typically have other identifiers such as network ID and bandwidth written on them. *Source*: Wikimedia Commons

it alerted online spammers to insecure corporate networks, potentially leading to illegal activity (Wearden, 2002). The phenomenon did, however, raise interesting questions around whether Wi-Fi is private property (Langley, 2003), particularly given that the connection is owned by the telecommunications company. Since then, the idea of being able to access Wi-Fi networks run by households has been formalized and commodified through networks such as Fon.

Community Wi-Fi as open infrastructure

How do we understand the contributions and innovations of community Wi-Fi networks? For a start, Wi-Fi communities share characteristics with free software communities. The latter evolved in response to corporations such as AT&T using freely developed and distributed software and locking it away in expensive licences. Developers began to write code, stipulating that it should be left 'open' and that others could contribute to it. Open source software and community Wi-Fi networks both involve groups of people working on technologies that possess a degree of openness. In the case of free and open source software, this is an openness to allowing others to access the source code and develop it further, or adapt it into other applications. Yochai Benkler (2006) describes what this means in relation to Wi-Fi:

> Throughout most of the twentieth century, wireless communications combined high-cost capital goods (radio transmitters and antennae towers) with cheaper consumer goods (radio receivers), using regulated proprietary infrastructure, to deliver a finished good of wireless communications on an industrial model. Now WiFi is marking the possibility of an inversion of the capital structure of wireless communication. We see end-user equipment manufacturers like Intel, Cisco, and others producing and selling radio 'transceivers' that are shareable goods. By using ad hoc mesh networking techniques, some early versions of which are already being deployed, these transceivers allow their

individual owners to cooperate and coprovision their own wireless communications network, without depending on any cable carrier or other wired provider as a carrier of last resort ... We must ask what, if any, are the implications of the emergence of a feasible, sustainable model of a commons-based physical infrastructure for the first and last mile of the communications environment, in terms of individual autonomy? (2006, pp. 153–4)

Wi-Fi is an internet infrastructure that anyone can participate in constructing, and where people may choose to become service providers for others under the conditions of community association ('coprovision' as Benkler (2006) calls it).

The technologies that community Wi-Fi groups develop can involve hardware as well as software, or they may use existing technologies but seek to disseminate these by raising awareness and teaching others how to run the equipment. These are also values-based communities whereby the shared value is one of openness – the belief that technologies should enable participation and be free from corporate enclosure (Benkler, 2006). The values that they bring to the technologies are therefore constituted in those technologies. De Filippi and Tréguer (2015) describe community networks as a 'subversive rationalization' (p. 2) of last-mile network infrastructure. Values include 'commitment of grassroots community networks to promote users' autonomy and fundamental rights to communication and privacy', as well as net neutrality (p. 5).

These groups are not so different from community associations that have produced debate and raised awareness of issues, cultures, and needs. As anthropologist Chris Kelty (2008) points out, software is speech. Through the creation of open technologies, communities are contributing ideas to what the internet should be. Kelty calls free software communities 'recursive publics': publics in the sense of a public sphere that generates information and decisions through the dynamics of deliberation. He considers them to be a 'public' as they are about making things public (source code), have distinct ideologies and beliefs, and operate outside of any formal organizational structure. Mechanisms such as GitHub, for instance, allow for decision-making to occur by providing a site for collaboration on code projects including version control. Meet-ups provide a face-to-face or face-to-screen interaction with others in the group. Ultimately, the code is a form of language that has the 'last word' – only by finding errors in code is a particular direction refuted and altered.

Recursive publics would not exist if the code had not been open source, as it is through the code that they come into existence. In Kelty's words: 'Recursive publics are publics concerned with the ability to build, control, modify, and maintain the infrastructure that allows them to come into being in the first place and which, in turn, constitutes their everyday practical commitments and the identities of the participants as creative and autonomous individuals' (Kelty, 2008, p. 7).

In the case of community Wi-Fi networks, there are different conditions that hold them together. As they exist only if enough people decide to join and extend the network, diffusion through a community is essential to their mission. The Wi-Fi groups above are (or were) committed to altering the hardware and software through 'hacking', to the point where it becomes more accessible and open. Those who use the network do not necessarily have the same commitment to openness (see Powell, 2012).

Community innovation

Another way to understand the importance of community networks is to study them as a process of innovation. In their study of Wireless Leiden, Van Oost, Verhaegh, and Oudshoorn (2009) found that, because the network was designed to meet public needs, there was an openness to information-sharing that would not otherwise have existed. While the two radio engineers had established a long-distance connection using expensive antennas between their homes prior to Wireless Leiden, the community project needed to be low cost, and accessible for others to install and use. The community nature of the project therefore generated different outcomes than would have been achieved as a closed or private endeavour. Between 2002 and 2004, the network expanded from 4 nodes with 12 volunteers to 50 nodes, 80 volunteers, and 2,000 users.

Innovation theorist Eric von Hippel has identified what he calls a 'free innovation paradigm'. In his definition, free innovation involves 'innovations developed and given away by consumers as a "free good" with resulting improvements in social welfare' (von Hippel, 2017, p. 1). Free software as described above fits within this definition, but the focus is on new ideas and products that arise from the activity, as opposed to political formations. Free innovation can compete with commercial innovation, such as Apache competing with Microsoft. It can also complement commercial innovation and spill over to become the basis for a commercial product (von Hippel uses mountain bikes as an example of this, whereby innovations from users are integrated into products). It can also work the other way in that commercial producers might also provide resources to support free innovations, as was the case with Linksys.

With community Wi-Fi, there is typically no individual lead user innovator who initiated and shaped the development of the Wi-Fi network, but, rather, diverse roles that come together. Van Oost, Verhaegh, and Oudshoorn (2009) call this 'community innovation' – innovation that is reliant on coordination of various elements and where the technical innovation and social group formation co-evolve. For instance, in the case of Wireless Leiden, new volunteer roles were created to ensure that nodes were maintained. While Wireless Leiden was initiated by a technical expert, it developed and grew through the cooperative efforts of many. In this instance, that included volunteer lawyers who drew up contracts to ensure the volunteers did not use hardware or knowledge gained through their involvement for profit.

Innovations have also emerged from competitions at events where Wi-Fi communities congregate. During a DEF CON hacking shoutout, contestants would use a pair of computers, establish a Wi-Fi connection on each, and see how far apart they could place the computers and still maintain a connection without using amplifiers. Technologist Brian Waldern presented his winning entry to the NYCwireless blog, describing how he and his team used a hotel room, a rental SUV, a base access point, two omni-directional antennas, two directional antennas, a remote client, Wi-Fi cards, pigtails and cables, and tripods and masts to produce a signal 10 miles out (4DI, 2003). They fared better than subsequent teams who built antennas from Hormel Chili cans and Pringles chips containers, reaching only 0.82 miles (*Wired*, 2004).

Figure 4.3 'Another view of Cantenna II'. *Source*: Flickr/lungstruck licensed under Creative Commons CC BY-SA 2.0 https://creativecommons.org/licenses/by-sa/2.0.

Various summits and conventions contributed to the development of community Wi-Fi and helped to sustain interest in community information infrastructures by bringing together technologists, community organizers, mapping groups, and policy activists (see Forlano, 2008, pp. 78–81).

Community informatics

A strand of academic research has been dedicated to observing community internet infrastructures, including Wi-Fi networks. A significant body of work came out of Canadian Research Alliance for Community Innovation and Networking (CRACIN), which was formed to look at 'how those at the economic and geographic peripheries are responding to the risks and opportunities presented by the new information economy', with a focus on what they called community informatics (Gurstein, 2012, p. 36). Michael Gurstein defined community informatics as 'the application of information communications technology (ICT) to enable and empower community processes' (Gurstein, 2007, p. 11). Community informatics, while concerned with the digital divide, is more specific in that it centres on the notion of 'effective use', and the principle that people need more than access in order to experience the benefits of internet use. For Gurstein, this included a recognition that individual and family well-being are contingent on the local environment. If effective use of the internet is inherently contextual, then a one-size-fits-all approach to internet provision may fail to account for specific capacities and needs. Studies of community informatics therefore focus on local community-run networks and services, and applications that are of benefit to a defined area, including content provided to a community either on the internet or through intranet-type arrangements.

Gurstein differentiates community informatics from ICT for development (ICT4D) approaches that entail

NGOs transferring skills and technologies to underserved populations. Community informatics projects, by contrast, commence when a community defines its needs and self-organizes to provide a network or internet facility. Community informatics is not restricted to Wi-Fi – it can consist of telecentres or 'wired' internet services (a quintessential example being K-net, an Indigenous-run internet network in northern Canada). As discussed in Chapter 2, Wi-Fi was designed to be self-provided, making it a convenient option for community-run networks, and a cheap option in areas where cables are prohibitively expensive.

Some communities have experimented with using Wi-Fi to share local content without connecting to the public internet. These systems connect the user to content available off a local server, which reduces costs by eliminating the backbone service. In areas where the backbone is provided by a satellite subscription, this reduces the cost of providing the hotspot and means that high-bandwidth content, including video, can be accessed without that subscription's data plan running out. Such systems also help to resolve the problem of internet affordability for low-income communities, as residents are able to access the content that they might otherwise choose to use mobile data for. Examples of this include Aboriginal media organization NG Media, located in remote Australia, which has been using Wi-Fi to stream content from a local content server within communities, in addition to its broadcast services (IRCA, 2017). Such services tend to work in areas that experience forms of digital exclusion (Echániz and López Pezé, 2018).

Mesh networks

In a mesh network, Wi-Fi access points are interconnected to form their own network, whereby each node relays data for other users of the network, not just the node owner (Navarro, Maccari, and Lo Cigno, 2018). Mesh networks are constructed so that all nodes cooperate non-hierarchically in the distribution of data to the mutual benefit of all within the network, so that resilience of the network increases as more nodes are added. If one node is connected to the internet, it can share it with others in the network. When a node joins or leaves a network, others reconfigure to ensure that everyone has connectivity. The routing protocol automatically selects routes between devices until it reaches its destination (multi-hop communication).

Mesh systems have been used in community networks, social movements, and for local non-commercial internet services. For instance, NYC Mesh had over 200 member-nodes by the end of 2018. The network has managed to avoid going through an ISP by establishing its own internet exchange point (at an establishment cost of $5,000, and $1,000 a month to run). Those who choose to join NYC Mesh pay some establishment equipment and installation costs, as well as an optional monthly donation. In addition, NYC does not collect personal data and is a neutral network (meaning it does not throttle data). Another example of mesh networking is that used in the One Laptop Per Child programme, which provides robust devices to children in low-income areas and developing countries. The devices are designed to connect to each other, allowing teachers and students to share content and participate in collaborative tasks.

Community mesh networks have also proven useful for temporary communication needs. For instance, community networking pioneer Sascha Meinrath and the Open Technology Initiative developed the Commotion Wireless system, a portable mesh network system that connects mobile devices. The technology was tested by the Occupy movement in 2011 to communicate by connecting people's devices, spreading internet through the camp off one office's donated connection (Singel, 2011). The same technology was used in the aftermath of Hurricane Sandy in 2012 to spread connectivity when other services were down.

In 2014, protesters in Hong Kong used Firechat, a now discontinued chatroom and messaging system that used Bluetooth and Wi-Fi radio within mobile devices to connect to other devices without needing an internet connection. Developers of the software, Open Garden, said that millions of chats were active during the 2014 Hong Kong protests, with hundreds of thousands of downloads of the app (Cohen, 2014). Another messaging app designed specifically for activists and journalists is Briar, which can sync via Bluetooth or Wi-Fi and via the Tor network when the internet is up. The encrypted connections ensure that information can be shared without being subject to surveillance or censorship.

While mesh networks have attracted interest for their use in protests and social movements, often these are used in combination with other technologies to ensure essential communication. For instance, Spanish authorities imposed a range of restrictions on citizens to prevent the Catalonian referendum on independence in 2017, including suspending laws related to referendum, and police raids of IT centres and government departments, and blocking websites (Poblet, 2018). Activists posted information to the open source developer repository, GitHub, and used Telegram messaging only as a last resort. When public schools' internet connections were shut down, prohibiting many polling stations from accessing the referendum application, roughly 100 stations used guifi.net to bypass the internet outage (2018).

The limits of community-based Wi-Fi networks

While community networks have motivated communities, provided free access to the internet, and promoted local arts and services, they are not entirely separate from the telecommunications market. With a few exceptions (such as NYC Mesh), community Wi-Fi networks typically rely on a contractual arrangement with an internet provider and on the infrastructure that has been built through iterative for-profit and government initiatives. Community networks often also involve direct collaboration with different kinds of organizations which can influence the design of the networks and where they are located. Some commence as community Wi-Fi and then become part of public (government-supported) networks. Île Sans Fil, for instance, was renamed Zone Access Public Montréal in 2016, and its network has been extended through a partnership with the City of Montréal.

In their study of the thriving free Wi-Fi ecology in Austin, Texas, Fuentes-Bautista and Inagaki (2006) found that there were multiple partnerships involved between industry, community-based groups, academic and research institutions, and local government. These relationships formed for reasons of funding, technical equipment, visioning (ideas), and volunteer cooperation. However, they also found that the community and government Wi-Fi were more likely than private providers to recognize Wi-Fi as a resource that all citizens should be able to access. Private Wi-Fi services were more focused on venues, and had not broadened to cover non-profit organizations or low-income neighbourhoods. Private providers worked on the assumption that users would engage in economic transactions in places where free Wi-Fi was located.

The Guifi.net Foundation has developed models for municipal governments that would help establish what they call 'commons infrastructures'. The model is similar to that which led to public access television stations in the United States, whereby private companies wishing to lay cables on public lands must assign capacity for public and community use. Community network researchers Navarro, Maccari, and Lo Cigno (2018) write that community networks also have to make trade-offs between open source and proprietary technologies; stable network management and planning software tools are not available as open source, thereby requiring the use of proprietary protocols. Hardware, with some notable exceptions, is closed source, 'with protected intellectual property and product secrets held by the industry for hardware such as radio boards, radio firmware, device drivers and programming interfaces' (2018, p. 18).

Even though community Wi-Fi networks may be embedded in various infrastructures and contractual arrangements, this does not necessarily reduce their power or significance. One community networking pioneer, Andrew Garton, describes the lasting legacy of community that emerged from one short-lived experiment. TS Wireless attempted to use wireless infrastructures to provide independent local music across a neighbourhood, but was ultimately thwarted by music royalty collection systems. Nonetheless, there were lasting outcomes:

> [Community] was there, and still exists, through the network of software developers, web coders and designers, passionate wardrivers and NetStumbler aficionados. These are the people we rarely see, who had created some of the more experimental and wildly innovative networks of their time. We had worked in remote villages in Africa, Southeast Asia and Indochina. We provided training and advice to regional and rural telecentres in Australia, including creating some of the earliest websites for community groups, non-government organisations and small businesses in the country. We had, at personal expense, established the means for anyone interested to learn about wireless networks and how they may foster community, belonging, nurturing curiosity and innovation where it may otherwise languish. Our experiment may have failed on paper, but it succeeded in bringing together a group of people who shared an experience that we had all grown from, and the ripple effect of that gathering expands still. (Garton, 2018, p. 71)

Finally, community networks are still limited by spectrum regulation. For instance, radio spectrum that is reserved for broadcast purposes is often vacant in rural areas because the commercial licence holder does not consider it to be a viable market. Such spectrum could be used for community networks, but is not, due to spectrum regulation. As mentioned in Chapter 2, dynamic use of TV 'whitespaces' spectrum – that which is leftover in bands reserved for digital television transmission – is now beginning to be explored. While NYC Mesh was able to establish its own internet exchange point, regulatory barriers in many countries make it prohibitive for small groups to take this approach. For instance, the Australian government mandates that ISPs collect metadata of users, creating overheads that would prohibit grassroots networks. In Germany between 2010 and 2017, the so-called Störerhaftung law resulted in Wi-Fi network owners being liable for the activities of those who used the network. In response, German community network Freifunk created what they called the Freifunk Freedom Fighter Box – a VPN service that connected to a Swedish internet provider (Lehnardt, 2012). Other community networks, including guifi.net, have developed terms that users must agree to (the FONN Compact in guifi.net's case), which shift the responsibility and risk back to users. Other than such self-regulatory measures, Federica Giovanella suggests that users could conduct informal monitoring of networks and report suspicious conduct or implement filters on gateway nodes (Giovanella, 2016, p. 119).

Community governance and Wi-Fi

As discussed in Chapter 2, Wi-Fi can be a cost-effective means of providing internet to areas where laying fibre-optic cable is deemed too expensive by private companies and public providers. Settlements in very remote areas, such as Australia's remote Indigenous communities, sometimes only receive internet via satellite, sharing that connection via Wi-Fi networks. Public and community Wi-Fi operators have also established hotspots in areas where there are fibre-optic and mobile broadband services in order to provide internet access for those who cannot afford these services, and in situations of temporary need (for instance, when a person has run out of mobile data credit and needs to access services). Although

Wi-Fi is often constrained by reach and daily limits on data downloads, it is often sufficient for low-bandwidth online tasks such as banking and messaging. The arrival of public or community Wi-Fi hotspots in remote areas has also given rise to questions about who controls these services and what responsibilities come with internet provision.

For example, some Aboriginal organizations and youth services working in remote communities have advocated for Wi-Fi instead of commercially provided mobile broadband as it provides the community with some control over internet access. In these instances, communities are determining aspects of internet use even where the infrastructure is provided through a commercial company or public agency. Wi-Fi makes it relatively easy to set and enforce rules through actions such as installing filters, controlling when internet is available, and turning it off if it is misused. At a forum to discuss internet needs, led by the peak body for Indigenous media and communications, First Nations Media Australia (FNMA), participants identified a need for a 'Locally designed and delivered program delivery model by community-controlled organisations including local control over internet/WiFi sharing for safety and cultural authority' (FNMA, 2019).

In 2014, an organization called Ethos Global conducted consultation work with seventeen communities in Arnhem Land (in Australia's north) for the Department of the Prime Minister and Cabinet, exploring various aspects of Wi-Fi provision, including determining community members' digital abilities, their experience of the internet, and internet speeds. The communities all had relatively new Wi-Fi services that were provided by a company called APN, under a contract with the government to provide basic telephony services to very small communities. Ethos Global's investigation found that the majority of people had little experience of the internet, although there were devices present in all locations that could be used to access it. In addition, 'All communities specifically asked for the ability to turn the Wi-Fi signal off and on at their request to assist with managing the usage from time to time. Mostly this was with particular reference to over-use or abuse by young people' (Ethos Global, 2014, p. 7).

Communities can also be affected by where a Wi-Fi hotspot is placed, and seek to coordinate internet access spatially using Wi-Fi. Locating a hotspot in an area that is dominated by men will disadvantage women, particularly where men's and women's spaces are clearly delineated (see Rennie et al., 2016). Some communities also choose to control the times that Wi-Fi services are available, automating the Wi-Fi to turn off at night so as to ensure that young people are not on the streets looking for connectivity. At least one community chooses to turn it off on Sunday to encourage people to spend time with their families (Rennie, Yunkaporta, and Holcombe-James, 2019). Some choose to turn the community Wi-Fi off during school hours as it helps to prevent truancy.

Governing internet connectivity is also a tactic for diffusing conflict. Escalating tensions (either between two groups within a community or across different communities) can have many causes, but are frequently said to be fuelled by social media communication (Vaarzon-Morel, 2014). According to a worker in an organization that provides Wi-Fi in remote communities, a man took an axe to a mobile phone tower in remote Western Australia as he felt it was partly to blame for a rise in conflict in his community. The story evokes the powerlessness that can be experienced in relation to internet provision; the 'always on' nature of the mobile network meant that the community could not determine for themselves the terms on which internet and social media were made available.

Indigenous governance systems that have existed since pre-settler times include methods for handling disputes. Law scholar Larissa Behrendt writes that these traditionally included adjudication through meetings, through public shouting or yelling to air the issue, and through exile or temporary exile. More violent methods of punishment were conducted in a controlled manner (Behrendt, 1995, p. 21). While these forms of dispute resolution are still used today, they are carried out face to face. Conflict that appears or is perpetuated on social media can be difficult for Elders in communities to control because they are not necessarily in the right place to intervene, or not using the platform themselves. Imposing terms on Wi-Fi connections can therefore be seen as a continuation of existing governance, whereby those with authority can diffuse tensions by managing where and when people use the internet. Although filters are another way to do this, social media sites such as Facebook and YouTube have proven difficult to filter out via routers as they regularly change their IP address.

Having listened to these concerns, one of the companies that provides Wi-Fi in remote areas, Easyweb Digital, implemented a system whereby Easyweb staff can control a Wi-Fi network from their office in Melbourne. When they receive a call from a person in the community, typically someone charged with looking after the network, they can switch off the connection and turn it back on when directed. In 2019, Easyweb stated they had only used this facility a couple of times, indicating that while these capabilities are desired by older members of the community, they are rarely used in practice (Sacchero, in Rennie, Yunkaporta and Holcombe-James, 2019).

Conclusion

The adaptability of Wi-Fi components and the relatively low cost of establishing Wi-Fi networks has attracted community networking pioneers. Community Wi-Fi networks contain some or all of the following characteristics: they are dependent on the interests and skills of a small group of experts; they may be aligned with other community values and informational needs; they can possess a public good motivation in that they seek to provide internet services to those who might otherwise not have access; or they seek to advance the open source and free software movements. In reality, such networks are mostly entangled with private or municipal services. They nonetheless fulfil important functions, particularly where internet affordability is an issue or in remote contexts where wired services are not an option. In addition, Wi-Fi provides the means for communities to manage and control aspects of internet connectivity, such as imposing filters and limiting access at certain times. These governance dimensions suggest that when local communities have agency over their internet connectivity, they can shape it to their own needs. Community Wi-Fi is therefore not only about access to the internet – it can also be a means to turn it off.

5
City

In 2011, three Oslo-based researchers, Einar Sneve Martinussen, Jørn Knutsen, and Timo Arnall, developed a project to give visual form to otherwise invisible Wi-Fi network signals. In their project, entitled 'Immaterials: Light Painting WiFi', the team used a 'WiFi measuring rod' to survey and map urban Wi-Fi coverage (Martinussen, 2011). The rod, a 4-metre-tall pole, incorporated a mobile Wi-Fi antenna and a band of lights that spanned the length of the pole. With the help of RSSI (received signal strength indication) sensors, the rod measured surrounding Wi-Fi signal strength and converted this signal strength into pulses of light. The height of the pulsing lights on the rod was a reflection of signal strength: a fully lit rod meant a strong signal; a partially lit rod reflected a weaker signal. The team slowly walked Oslo's city streets at night, carrying the Wi-Fi measuring rod. Time-lapse photographs of these night walks produced 'light paintings' that created an illuminated topographical map of sorts, depicting variable urban Wi-Fi coverage in the city (see Figure 5.1).

In reporting on this project, Martinussen (2011) notes how ubiquitous city Wi-Fi has become, yet how fragmented Wi-Fi network coverage tends to be; how the behaviour of signals depend on where the network is located, and how the surrounding cityscape has developed and evolved; and how Wi-Fi networks are difficult to contain, and that 'architectural forms, building materials and the urban landscape shape how networks spread into the city' (2011) – when it comes to Wi-Fi networks, Adrian Mackenzie (2006a) notes, data flows and overflows (see Figure 5.2).

Figure 5.1 Immaterials: Light Painting WiFi (2011). *Source*: Einar Sneve Martinussen, Jørn Knutsen, and Timo Arnall, The Oslo School of Architecture and Design.

Figure 5.2 'A WiFi network from an 1890s apartment building spilling into the street' (Martinussen, 2011). *Source*: Einar Sneve Martinussen, Jørn Knutsen, and Timo Arnall, The Oslo School of Architecture and Design.

The Immaterials project illustrates nicely a number of concerns that sit at the heart of the discussion of Wi-Fi and cities in this chapter. These include the way that Wi-Fi sits alongside and in-between other urban and telecommunications infrastructures, and how Wi-Fi remains 'highly mutable' (Mackenzie, 2010, p. 35). Wi-Fi saturates cities, yet does so in ways that can be variable and uneven. Moreover, in urban contexts, Wi-Fi tends to be characterized by, and subject to, instability of form and function, which renders it 'mutable, transitory, and malleable' (Cooper, 2002, p. 19). And yet Wi-Fi networks have reconfigured people, places, and information in cities in significant ways (Forlano, 2009). How this is so is still yet to be fully appreciated. One explanation for our lack of detailed understanding of the impacts and importance of Wi-Fi, as with quotidian mobile use, is that 'some of its qualities might be said to be obscured by its status as both an already routine feature of everyday life, and yet one which continues to be subject to rapid change' (Cooper, 2002, p. 19).

A key aim of this chapter, then, is to make our everyday, routinized engagements with city Wi-Fi strange again. We begin by taking a step back to consider the central place that communication technologies have long held and continue to hold in the history and development of cities, before situating Wi-Fi within this history. We then give consideration to what we regard as two factors fuelling the acceleration of city Wi-Fi network development and normalization of Wi-Fi use: the rise of Wi-Fi-enabled portable computing and mobile devices; and cafés as key sites for the provision of urban Wi-Fi access. The argument we develop in relation to cafés is that the role they play in supporting Wi-Fi access for urban occupants makes sense when we look back at the café through history, where it has always operated as an important public–private site of sociability and communication.

The emphasis then shifts in the second half of the chapter. Our examination of Wi-Fi here is anchored around the idea that city or municipal Wi-Fi can be understood productively as *a contested space of possibility*. This deceptively simple formulation provides a useful frame for making sense of: (a) the fraught nature of the politics and economics of ('free') public Wi-Fi; (b) the promise (fulfilled and unfulfilled) of municipal Wi-Fi, especially in relation to questions of equity and access; and (c) the place of Wi-Fi within visions of the connected or 'smart' city. Throughout the chapter, we also note various attempts, over the course of Wi-Fi's relatively short history, at 'platformizing' city-based Wi-Fi signals and networks.

Communication technologies and the city in history

Cities are, and have always been, places for, about, and products of communication (Gumpert and Drucker, 2008, pp. 195–6). In Lewis Mumford's (1979) influential book *The City in History*, he notes how, historically, communication was vital to the promotion of civic life within cities. In Mesopotamian times, for instance, citizens were called to assembly via the use of a drum; in medieval times, this function was served by the church bell (1979, p. 79). Communication reach also served as an important 'conditioning factor' (p. 79) in determining the scale of cities. 'The permissive size of a city', Mumford writes, is dependent upon (among other things) 'the velocity and effective range of communication' (p. 80). Thus, larger cities were only possible where they became 'the centre of a network of communications' (p. 80). Mumford's understanding of a 'communication system' is a deliberately expansive one, taking in alphabets, permanent records, waterways, and other forms of transport infrastructure, and, of course, more modern systems of telecommunication. It is for this reason that, in his historical account of urban development, Mumford suggests that 'more important, in the long run, than the wider distribution of the goods in the market' in contributing to the formation of large metropolises 'is the wider communication system that grew up along with it' (p. 88).

By the late nineteenth century, cities had begun to expand rapidly in geographic scale and density. These developments were occurring in tandem with equally rapid developments in communication technologies, which exerted both centrifugal and centripetal forces upon the city (McQuire, 2008, p. 16; see also Mattern, 2017, p. 25). On the one hand, urban and suburban expansion was facilitated by 'new communication technologies such as the telephone, which supported the coordination of spatially separated production and retail sites in the factory system' (McQuire, 2008, p. 16). And, on the other hand, the formation of central business districts (CBDs) and 'the logistics of office work demanded communication networks such as telephones capable of servicing multiple cells aggregated in monolithic structures such as the high-rise tower' (p. 16; see also Castells, 1989, p. 149).

In *Telecommunications and the City*, Stephen Graham and Simon Marvin chart this growth in communications systems in detail. Early in their book, they represent this growth in telecommunications through a loosely funnel-shaped diagram (see Graham and Marvin, 1996, p. 16) that begins with the emergence of the telegraph in 1875, and expands outwards as new communications technologies were developed and came online. By the mid-1980s and beyond (the wide end of the funnel), cities had come to be regarded as 'giant engines of communication' (Graham and Marvin, 1996, p. 6). This is largely due to advances in switching and transmission technologies that led to the development of four new, predominantly city-focused, telecommunications infrastructures: (1) 'wireless and mobile communications systems' that 'link telephones and computers by radio signals to fixed telephone networks' (Graham and Marvin, 1996, p. 20); (2) 'broadband cable networks' (p. 21); (3) 'a new generation of satellite infrastructures' (p. 21); and (4) 'microwave systems' (p. 23). McKenzie Wark refers to these infrastructures collectively as a kind of 'third nature': 'From the telegraph to telecommunications, a new geography has been overlayed on top of nature and second nature.... Second nature, which appears to us as the geography of cities and harbours and wool stores is progressively overlayed with a third nature of information flows, creating an information landscape which almost entirely covers the new territories' (Wark, 1994, p. 120).

A further way these overlays have been framed is as existing in 'hertzian space' (Dunne and Raby, 2001; Dunne, 2005). Hertzian space, as Dunne and Raby (2001, p. 12) conceive of it, is a 'medium for carrying information, an invisible alternative to wires and cables'. The communication technologies that exist in and flows that criss-cross hertzian space are woven together to make a landscape – a 'hertzian landscape' (Dunne and Raby, 2001, p. 20), or 'spectral geography' (p. 18) – that 'overlays and overflows topographical, geographical differences between points' (Mackenzie, 2006a, p. 141).

Wi-Fi, however, does not necessarily fit that well with these formulations of 'third nature' or of 'hertzian space'. This is for two reasons. First, within cities, Wi-Fi connectivity is often far from straightforward, involving all manner of dropout, and signal strength and connection issues.

This is a particular problem in dense urban built environments, where urban canyons and other barriers can interfere with telecommunications signals. Wi-Fi end-users have quickly learnt to adapt to this. For example, as Sung-Yueh Perng (2015) found in his study of Taiwanese Wi-Fi use, those living in Taipei developed a range of tactics and day-to-day practices, developed from accumulated past experience, for negotiating fluctuating levels of connectivity while moving about the city.

One of us witnessed a memorable example of these same connection issues firsthand during fieldwork in the Old Town of Tallinn, Estonia. We were sitting drinking a coffee in a café that was situated in a medieval building below street level, when a tourist couple entered and asked the waiter whether the café had free Wi-Fi. The waiter responded: 'Well, outside, yes, but inside it gets a little more complicated. There is Tallinn [municipal free] Wi-Fi out in the square [he pointed to Raekoja Plats, the main town square, which the café bordered]. Or, if you move into that room [pointing to an even lower room, that abutted the square], you might get better reception, including access to Tallinn Wi-Fi.'

In giving this advice, it was unclear whether the issue was really one of reception / signal strength, or a desire on the part of the waiter to offload tourist traffic from the café-provided Wi-Fi onto the municipal free Wi-Fi; our suspicion was that it was likely a bit of both of these things.

The second point is that, within cities, the 'invisible' coverage and information flows and overflows of data criss-crossing the urban landscape are in fact the product of a 'byzantine array' (Mattern, 2017, p. 37) of wired and wireless infrastructure that work collectively to beam 'imperceptible, but still very much material, waves at all that inhabits the streets below' (p. 37). Just as new technologies that permit communication across distances 'have tended to function as additions to, rather than replacement for, previous mobilities' (Morley, 2017, p. 101), so it is with Wi-Fi. The technologies that support Wi-Fi are built onto and into buildings, and interconnect with and rely on a plethora of other communications technologies and infrastructure. In this way, the information infrastructure of contemporary cities has come to constitute a 'heterogeneous mix' of wired and wireless access technologies that connect and integrate Wi-Fi or WLAN technologies into 3G, 4G, and 5G, as well as other wireline systems (Lehr and McKnight, 2003, p. 353). The result is that Wi-Fi networks, in combination with other technologies, form a 'patchwork quilt' (Bar and Galperin, 2004, p. 45) of wired and wireless urban information and communication infrastructure that is embedded within the city.

What do we mean, then, when we talk of city Wi-Fi? In essence, Wi-Fi is a radio-frequency-based system (employing the 802.11 standard) that permits an enabled device (such as a laptop, tablet, or smartphone) to connect to the internet over a wireless local area network (WLAN) through an access point. These access points might be situated within domestic homes, offices, or other commercial and civic and educational settings. Because of their limited range, small networks are usually referred to as 'hotspots'. Clusters of access points or joined locations that provide access to a Wi-Fi network are often referred to as a 'hotzone'. A hotzone might cover a restricted area, such as a specific retail precinct, or it might extend to cover a whole city. At larger urban scales, connectivity, as we discussed in Chapter 4, often occurs through mesh networks. Whereas traditional networks rely on a small number of wired access points or wireless hotspots to connect users, with a wireless mesh network a network connection is distributed among multiple wireless mesh 'nodes' that interact with each other to share the network connection across much larger areas (see Mackenzie, 2007, p. 96). In this chapter, we primarily focus on individual users of localized 'hotspots', as well as, in the second half, on larger 'free' municipal services of varying scales of reach and complexity.

Against this backdrop of urban and technological development, Wi-Fi, as the patchwork quilt metaphor suggests, tends to be regarded less as a replacement for wired broadband access than as a complementary technology – Wi-Fi 'has always been accompanied by a range of signal sources' (Perng, 2015, p. 289). The aforementioned connectivity issues notwithstanding, Wi-Fi also provides coverage for places and for users that other networks have difficulty reaching (Hayes and Lemstra, 2009, p. 68). As we discussed at length in Chapter 2, Wi-Fi establishes parallel access infrastructures while also populating gaps between existing urban infrastructures (Mackenzie, 2010, p. 35) – such as in the way that it sits, interstitially, bridging wired and mobile wireless technologies (Lehr and McKnight, 2003). In this sense, Wi-Fi can be seen to 'lend extraordinary versatility' (Castells, 1989, p. 149) to the existing urban telecommunications ecosystem.

Not only do urban Wi-Fi networks add flexibility to the telecommunications ecosystem of contemporary cities, they have also come to be regarded – and commercially exploited – as a vital location infrastructure in their own right. In the early 2000s, for example, US company Skyhook Wireless (and, subsequently, Google) developed systems for calculating location positioning from existing Wi-Fi signals rather than from cellular towers or by using GPS; Skyhook Wireless's Wi-Fi positioning software – an early attempt at 'platformizing' Wi-Fi – was incorporated into first-generation iPhones (Wilken, 2019; Torrens, 2008). (These same Wi-Fi positioning techniques have since found a range of applications, including in Japan in civic computing where it is used for differentiating occupied from unoccupied housing (Konomi et al., 2017)).

Having provided this overview of Wi-Fi's somewhat complicated position within the urban telecommunications ecosystem, we give consideration over successive sections to two key things – one a set of relatively recent technological developments, the other a site or setting with a long urban history – that are regarded as having played a crucial role in driving the development and normalization of city Wi-Fi: the emergence of laptop computers and mobile devices; and cafés or coffee shops.

Device portability and the diffusion of city Wi-Fi

The rapid diffusion of Wi-Fi within cities, and with it the growing importance of Wi-Fi networks to city life, occurred in direct conjunction with developments in battery-powered portable or laptop computing.

Back in 1986, battery-powered portable computers commanded just 2 per cent of the global computing market (Lap-Top Computers, 1987). Apple was the first computer manufacturer to produce a laptop – the iBook, released in 1999 – with a Wi-Fi radio chip inside (Links, 2003, pp. 105–13). By the early 2000s, laptop computers – most, by then, Wi-Fi-enabled – still comprised less than a quarter of personal computers shipped (around 16 per cent) (Kanellos, 2009). This was perhaps due to the fact, as Adrian Mackenzie (2006b, p. 783) notes, that 'the idea that data would move in concert with people [was] still novel' at that time. This situation, however, was soon to change. 'What emerged from 2001 onwards', Gerard Goggin (2006, p. 174) writes, 'was the growth of wireless, broadband access through Wi-Fi-enabled laptops', following rapid integration of wireless-LAN functionality into various computing and networking products (Hayes and Lemstra, 2009, p. 66). By May 2005, US laptop sales surpassed desktop sales (Singer, 2005); internationally, this milestone occurred three years later, with 38.6 million laptops sold, compared with 38.5 million desktops (IHS Markit, 2008).

In addition to laptops, a host of portable devices with Wi-Fi connectivity were also released, including personal digital assistants (PDAs) in the early 2000s, such as HP's Ipaq range (Arar, 2003), as well as gaming consoles, including the Nintendo DS in 2004 and the Sony PlayStation Portable in 2005 (Goggin, 2011, p. 110). In 2005, Nintendo also 'launched its own Wi-Fi Connection Service, and then also offered a Wi-Fi USB connector, through which a number of users could play' (p. 110).

With the emergence of portable computing, Wi-Fi moved beyond the home and the office to be deployed in a range of wider contexts, including in offering connectivity to business and other travellers in hotels, airports, and train stations, within outdoor public spaces such as urban parks and city squares (Forlano, 2008; Hampton, Livio, and Sessions

Goulet, 2010), and in a range of privately controlled public spaces, such as cafés, bars, and restaurants (Bar and Galperin, 2004; Goggin, 2011, p. 32).

Further rapid diffusion of Wi-Fi in urban contexts, and growing reliance on Wi-Fi hotspots and networks, came with the arrival of Apple's iPhone and iPad and Google's Android operating system in 2007–8. Extraordinary growth in smartphone and tablet take-up and use followed, which led to the consolidation of the mobile internet (Goggin, 2011; Hjorth, Burgess, and Richardson, 2012). The rise of the smartphone – the 'iPhone moment', as Goggin (2011, p. 181) refers to it – is also significant in this context in that it accelerated the 'trend toward the crossover between Wi-Fi (wireless internet) and cellular mobile networks and devices', such that handsets, applications, and users could switch with relative ease between cellular and wireless networks (Wilken and Goggin, 2015, p. 6). For telecommunications providers, the appeal of Wi-Fi as an interstitial or complementary technology is that it permits mobile data offloading – whereby data originally targeted for cellular networks is 'offloaded' for delivery via other, complementary networks (Wikipedia, 2019c) – at certain times and in certain places (such as at airports or in stadiums) where demand is high.

With increased demand for urban Wi-Fi network access catering to urban dwellers and travellers came a range of market opportunities, on both the technology side and the consumer side. For instance, on the technology side, the early days of city Wi-Fi provision saw a multitude of tech companies spring up to provide Wi-Fi hotspots (e.g. Wayport, now AT&T Wi-Fi Services), Wi-Fi sharing services (e.g. Fon), and other Wi-Fi-related services, such as 'centralized subscription, billing, software, and technical support' (Sandvig, 2004, p. 583; Bar and Galperin, 2004) (e.g. Boingo Wireless, iPass), among other things. On the consumer side, fast food outlets and coffee shops and cafés (to be discussed further below) were among the first businesses to offer Wi-Fi to their customers. In the US, for example, Wi-Fi has been available in McDonald's stores at a cost since at least 2004 (Gural, 2010). Free Wi-Fi was introduced in their UK stores at the end of 2007 (Smithers, 2007), and in the US by the end of 2009 (Associated Press, 2009). Starbucks responded by introducing free Wi-Fi in the US from 1 July 2010 (Cain Miller, 2010), and this has since become a feature of Starbucks' stores worldwide.

In the decade that followed, subsequent urban Wi-Fi hotspot growth has been exponential, with over 1.7 million commercial hotspots and 74 million community access points in the USA alone since 2018. Wi-Fi also carries 67 per cent of mobile device traffic in the US, a figure that is slightly higher (83 per cent) in Japan (WBA, 2018, p. 9). And Cisco predicted that Wi-Fi, as a percentage of global IP traffic, will grow from 41 per cent in 2016 to 46 per cent in 2021 (Cisco, 2016). Such growth is reflected in the increased estimates of the value of the Wi-Fi industry. Telecom Advisory Services, for instance, placed the global economic value of Wi-Fi at US$1.96 trillion in 2018, rising to US$3.47 trillion in 2023 (Telecom Advisory Services, 2018).

The point we can draw from these figures is that Wi-Fi is now widely regarded as a vital part of the telecommunications infrastructural mix within most cities. 'In the digital world we now live in', the Wireless Broadband Alliance (WBA, 2017) suggests, 'Wi-Fi is no longer an amenity – it has taken on the role of an indispensable utility.' Nowhere does this seem more evidently so than in city cafés.

Cafés and the social life of Wi-Fi

While portable computing has played an important role in the wider take-up of Wi-Fi, it is no exaggeration to say that cafés also form crucial sites in both the early adoption and continued success of Wi-Fi in urban contexts. In order to appreciate why this is the case, it is valuable to consider the unique place and significance of cafés in urban life.

Cafés, Sharon Kleinman (2006, p. 191) suggests, 'serve important social, psychological, and economic functions' within the life of the city. Indeed, cafés have even been described as constituting 'the very heart of urbanism today' (Stenseth, 2013, p. 24). According to Lyn Lofland's (1989b) influential formulation, urban life can be divided into three realms: private realms (private households), parochial realms (local neighbourhoods), and public realms. The last of these, she writes, is 'made up of public places or spaces ... that tend to be inhabited ... by persons who are strangers to one another and who "know" one another only in terms of occupancy or non-personal identity categories' (Lofland, 1989a, p. 19). The public realm has come to be regarded as 'a defining characteristic of city life' (Montgomery, 1997, p. 86). A similarly influential and related distinction has been drawn by Ray Oldenburg (1989), who differentiates between one's 'first place' (home), one's 'second place' (work), and what he calls 'third places', which loosely correlate with Lofland's understanding of public realms, and which Oldenburg regards as vital to the social life of cities. Cafés are said to form vital 'third places' (Oldenburg, 2013, 1989), key sites within the public realm in which public social life can take place, and which have long worked to foster civil society, democracy, and civic engagement, and in establishing 'communities of place' (Kleinman, 2006, p. 191; Doyle, 2015).

Cafés thus form a social institution of immense importance and with a long history. The café tradition stretches back to the seventeenth and eighteenth centuries, with its roots in the salon culture of the French Republic of Letters (Goodman, 1994) and in the London coffee-houses (Ellis, 2004). As scholars such as Jürgen Habermas (1989) and Richard Sennett (1974) have argued, the London coffee-houses – such as Jonathan's Coffee-House, which stood on the original site of the London Stock Exchange – played an important

– if contested (Ellis, 2004; Laurier and Philo, 2007) – role in fostering the formation of a democratic public sphere. These coffee-houses were numerous. Dispersed throughout the West End and other parts of London, they were each quite distinct in flavour and in terms of the composition of the social groups that met there, with each differentiated, to use Lofland's language, according to micro-localities associated with the 'parochial realm' of different neighbourhoods, just as urban cafés are today: 'Merchants, insurance agents and brokers met at Jonathan's and Garraway's coffee-houses in Exchange Alley ... For wits and poets an important concentration of coffee-houses emerged in Russell Street, a broad street leading off the crowded piazza of Covent Garden, close to the theatres' (Ellis, 2004, p. 150).

What is regarded as historically striking about London's coffee-houses is, firstly, that they provided an important 'third space' – a place for social interaction and public social life – in which strangers could gather and mingle freely, and where speech was encouraged. Secondly, just as the Republic of Letters developed around epistolary media (Goodman, 1994), London's coffee-houses formed vital sites for both the production and consumption of news media (as well as giving impetus to 'the early postal development of London, which in time led to the organized delivery of letters, and the distribution of newspapers' (Lillywhite, 1963: 18–19)). As Eric Laurier and Chris Philo explain: 'The periodicals, *The Tatler* and *The Spectator*, were born within this society, reflecting its concerns, and expressly giving the impression of being written *from* coffee-house tables after coffee-house discussions. At the same time, they circulated around the coffee-houses, being bought there, often read there and then commonly the subject of debate there' (Laurier and Philo, 2007, p. 263). In short, London's coffee-houses formed key spaces of sociability and communication.

Cafés, as descendants of London's coffee-houses, and especially as they developed from the nineteenth century onwards in Europe and elsewhere, have continued to fulfil these functions, playing 'a crucial historical role in the formation of modern sociability and textual [production and] circulation' (Manning, 2013, p. 46) and communication. With respect to the first of these (sociability), cafés came to form a key locus for, and gathering place of, the European intelligentsia throughout the nineteenth and early twentieth centuries (see Rittner, Haine, and Jackson, 2016; Kusiak and Kacperski, 2012), as well as for the working class (Haine, 1999). With respect to the latter (cultural production and communication), Anthony McCosker and Rowan Wilken (2012) write that 'the design features of the café, the form and layout of the tables and the space made available for individual habituation, the din of social encounters, and even the stimulating influences on the body of the coffee served there, can be seen to act as enablers of communication and creativity'.

To illustrate, they give the example of Polish composer Krzysztof Penderecki, who, in the early 1960s, preferred the din of the 'third space' of Krakow's café Jama Michalika over the distractions of his 'first space' (a small apartment) or his 'second space' (the university music school) to compose. The strictures of the tiny café tables even led him to invent a special music scoring system for capturing in graphic notational form the fifty-two instruments that featured in his piece, *Threnody for the Victims of Hiroshima* (1960) (McCosker and Wilken, 2012).

This same dual role – a space for 'convivial sociability' (Haine, 1999) and for the production and consumption of communication content – continues into the present, and the age of Wi-Fi-equipped contemporary cafés (*Economist*, 2003), which operate at the 'nexus of face-to-face and mediated interactions' (Kleinman, 2006, p. 191).

One key shift that has occurred within cafés is that towards a 'more isolated, individualistic *habitué*' (McCosker and Wilken, 2012). Modern-day cafés provide important public–private spaces for those seeking physical co-presence with others; while this may involve socializing with friends or colleagues, it also, importantly, involves being alone in the company of others (Walters and Broom, 2013, p. 186). W. Scott Haine (1999, pp. 150–78) coins the term 'intimate anonymity' to describe the unique forms of sociability that now occur within cafés (in Japan, for example, it has been noted that cafés fulfil an important function in providing spaces of solitude for temporary escape from the strictures of modern Japanese life (see White, 2012, ch. 7)). Wi-Fi has been important in this regard, in providing a point of focus for many contemporary café dwellers who also come to enjoy coffee and 'intimate anonymity' in the company of co-present strangers. In this way, cafés continue to constitute complicated sites of interpersonal communication (with a strong gestural economy and detailed repertoire of glances, even if spoken interactions between co-present strangers are now largely reduced to phatic communication), as others have documented in detail (Laurier, 2008a, 2008b; Laurier and Philo, 2006a, 2006b; Laurier, Whyte, and Buckner, 2001; Tjora, 2013). In addition, cafés, as others have also observed of Wi-Fi-equipped public parks (Forlano, 2008; Hampton, Livio, and Sessions Goulet, 2010), now also constitute complicated sites of Wi-Fi-enabled computer-mediated communication, as when laptop users undertake various forms of 'situational domestication' (Henriksen and Tjora, 2018) when appropriating – or 'place-staking' (Hampton and Gupta, 2008) – café space for work purposes.

And yet the provision of Wi-Fi – especially *free* Wi-Fi – has come to be regarded as something of a mixed blessing for café and coffee-shop proprietors. On the one hand, Wi-Fi can attract customers, build additional revenue (such as through purchases made whilst using Wi-Fi), and build customer appreciation and loyalty (see Figure 5.3). On the other hand, there are trade-offs to be made, including around dwell time and/versus the number of tables (covers) a café can turn over

Figure 5.3 Café doors in the Old Town, Tallinn, Estonia, with a sticker promoting Wi-Fi availability. *Source*: authors' own.

in a given period, the cost of providing Wi-Fi, and the cost of supplying power to customers (O'Brien, 2009). We see this ambivalence playing out in the Tallinn example from earlier in this chapter, where it was unclear from our observations whether the waiter wanted the newly arrived customers to use his Wi-Fi or the municipal free Wi-Fi. In addition to these economic and operational considerations, and despite the earlier suggestions that cafés continue to constitute complicated sites of interpersonal and computer-mediated communication, there is also a view that Wi-Fi use can lead to a diminishment in the forms of rich social interaction that cafés as third spaces are seen to permit (Metz, 2017).

For those café proprietors who do offer Wi-Fi within their establishments, there are three main ways that this service tends to be set up and maintained. The first involves the café proprietor (merchant) setting up a WLAN, with the router usually supplied by their internet service provider (ISP) or telecommunications company, and then navigating the various establishment steps – choosing an appropriate data plan, setting up a guest network, creating a 'captive portal' (the web page that contains terms of use, and grants users access to the network), and so on (Reddigari, 2019). The second approach is for merchants to engage the services of a specialist Wi-Fi provision firm – such as Purple WiFi, Cloud4Wi, Skyfii, Presence Orb, Airangel, Fon, Easyweb Digital, or Encapto (a spin-off from Easyweb Digital), to name just a few of the more prominent companies – to handle Wi-Fi installation and guest access for them. In many cases, these companies also compete with specialist firms (like Aislelabs Flow) in offering a suite of location-based marketing and analytics services that draw on the same Wi-Fi traffic data they facilitate access to. The third approach, which follows the same logic as the second but warrants its own discussion, is where big tech companies, most notably Facebook, provide support for small businesses in providing Wi-Fi guest access. In late 2013, for instance, Facebook partnered with Cisco's Enterprise Networking Group to roll-out a 'free' Wi-Fi service to any business in the United States that wished to use it (Hajela, 2013). Bearing the rather awkward title of 'Connected Mobile Experiences (CMX) for Facebook Wi-Fi', the arrangement was that merchants would use their own existing router and broadband subscription, which then integrated with the CMX software (Constine, 2013). In exchange for each business signing up to the service, Facebook provides the merchant 'with the aggregate ages, genders, and other demographic info of those who check in, but in an anonymized format without names attached' (2013). In supporting this initiative, Facebook hopes to give its mobile and laptop users an incentive to engage with the service, ideally by using federated identity management and accessing the internet through the Facebook portal. In addition, Facebook hopes to encourage more businesses to sign-up as a 'prelude to buying ads' (Tate, 2012), thereby extending the reach of its local search, recommendation, and deals services, with the aim of become a 'formidable player' (Van Grove, 2013) in local search and location-based analytics.

What is striking about these last two approaches to the provision of Wi-Fi is that both frame guest access as forms of 'digital enclosure' (Andrejevic, 2007). The logic of the digital enclosure, Mark Andrejevic (p. 304) argues, 'combines the spatial characteristics of land enclosure with the metaphorical process of information enclosure'. Whether it is Purple WiFi or Facebook, the model is one of facilitating guest access, and using the data generated via this access to leverage and promote their own analytics expertise. It stands as another interesting attempt at seeking to 'platformize' Wi-Fi, in the case of Purple WiFi and the other specialist firms (Purple describe their service as 'a market leading guest WiFi, analytics and engagement *platform*'), or to enrol Wi-Fi into existing platform logics, in Facebook's case.

In the remainder of this chapter, we turn to a consideration of how city Wi-Fi can be understood as a contested space of possibility. We trace how this idea is played out in relation to the politics (and political economy) of city and municipal Wi-Fi, in relation to questions of access and equity, and in relation to connected or smart cities.

The politics and economics of city Wi-Fi

Adrian Mackenzie remarks on how Wi-Fi tends to resist becoming a 'coherent object of expression' and, instead, is 'something dynamically generated within a flow of meanings' (Mackenzie, 2006b, p. 796). There is, he writes, 'a profusion of competing ideas of movement, space, access and regulation' that make Wi-Fi 'difficult to categorize or even to delineate as a discourse' (p. 796).

Wi-Fi's status as a 'multivalent object of representation' (p. 785) is especially evident within the literature on public and community Wi-Fi projects and policies. For instance, Alison Powell, in her analysis of two Canadian Wi-Fi initiatives, Fredericton's Fred-eZone and Montréal's Île Sans Fil projects, proposes that hotspot-based Wi-Fi networks be conceived of as 'public parks' in addition to 'public utilities' (Powell, 2009). The appeal of the public park metaphor, for Powell, is that it promotes the idea of Wi-Fi as a medium that should, in a networked community context, provide space for sociability, participation, play, dialogue, and activism. Meanwhile, Claudio Luis de Camargo Penteado and colleagues (2017) conclude, from their study of WiFi Livre SP in São Paulo, Brazil, that 'the provision of WiFi service should be seen as a public service not a commodity to be purchased on the market' (p. 312). In the Australian context, Goggin (2007) considers wireless technologies, after Benkler (see Chapter 4), as common pool resources, using this framing as a way of opening up Australian policy debates beyond a focus on telecommunications as exclusively public or private goods. And, finally, McShane, Wilson, and Meredith (2014) conceive of Wi-Fi as a form of civic infrastructure, a formulation that 'acknowledges the hybrid arrangements through which local-level infrastructure has typically been provided and managed in Australia, often involving contributions from government, community and business groups' (p. 131).

There is also considerable variability in how the promise of city Wi-Fi is understood. Beyond facilitating the movement of data and people, Wi-Fi has, for instance, been described as holding potential for democratizing technology access (and, with community-run networks, of affording 'a chance to dismantle monopolistic ownership of information infrastructures' (Mackenzie, 2006b, p. 788)), driving entrepreneurial activity (Potts, 2014), providing opportunities for the emergence of new social actors and new publics (Powell, 2008), and activating public spaces (Forlano, 2008), including in ways that promote the social good (Lambert, McQuire, and Papastergiadis, 2013, 2014). A number of these arguments also resonate with particular strands of media and communication scholarship that seek to highlight the social and other civic possibilities of urban media and communications infrastructures for city dwellers (see, for example, de Waal, 2014; McQuire, 2008; Carpentier, 2008).

But it is not just how public Wi-Fi is conceptualized that can be characterized as a contested space of possibility. The business arrangements of, and investment rationales for, municipal and city Wi-Fi can be viewed in similar terms.

Municipal or city-based wireless networks (as opposed to those run by specific businesses, such as cafés) have tended to adhere to one of a limited number of ownership models (Forlano, 2008; Bar and Park, 2006; Park and Lee, 2010). Forlano (2008, p. 83) summarizes these models as follows:

> *Privately owned networks*, such as the hotzone around Picadilly Circus in London that was run by London-based internet service provider (ISP) Broadreach (Mackenzie, 2006a, pp. 138–9), or the short-lived municipal wireless network, Google Wi-Fi (not to be confused with its more recent household mesh router product of the same name), which Google ran from 2010 to 2014 in Mountain View, California (Wikipedia, 2020b);
> *Public–private partnership networks*, the most common ownership model, especially in delivering 'free' municipal Wi-Fi;
> *Publicly owned networks*, such as the early and influential but now defunct St Cloud, Florida, free citywide Wi-Fi service (Vos, 2009), or Fredericton eZone in Canada (Middleton and Crow, 2008); and,
> *Community-owned networks*, such as Montréal's Île Sans Fil (Middleton and Crow, 2008), or NYCwireless, the non-profit organization that is best known for its work in deploying free Wi-Fi in Bryant Park and Tompkins

Square Park in New York City (see Chapter 4, and, for detailed discussion, see Forlano, 2008).

Heer et al. (2010) categorize these arrangements somewhat differently, suggesting they tend to be arranged in one of three main ways: (1) as *municipality-driven networks*, where the intent is to provide Wi-Fi to city staff, citizens and tourists in key areas, such as Paris Wi-Fi or Tallinn Wi-Fi; (2) as *provider-driven networks*, which may serve municipalities but whose business models and interests are geared around the companies providing the service, such as the well-documented but now discontinued Wireless Philadelphia project, which was initiated by ISP Earthlink (Heer et al., 2010, p. 589; Shaffer, 2007); and (3) *user-driven networks*, where private users form a Wi-Fi community in which they share internet access, such as the Air-Stream group in suburban Adelaide, Australia, studied by Jungnickel (2014).

Despite a great deal of hope for – and hype around – municipal and city-wide initiatives, Mackenzie aptly describes city Wi-Fi's fortunes throughout the 2000s as a 'precarious market adventure' (2006b, p. 786). This is a characterization that has been reinforced elsewhere. For example, Goggin (2007, p. 120) has suggested that 'the underlying problem is the economics of commercial Wi-Fi, with a business case difficult to pin down' (see also Bar and Park, 2006). Indeed, Jason Potts (2014) found no clear economic basis for the introduction of municipal Wi-Fi: 'municipal WiFi is not a natural monopoly, it is not subject to market failure; public provision would likely distort existing markets and devalue existing investments'. Catherine Middleton (2007) also expresses doubts about the wisdom of municipal investment in wireless networks, suggesting these investments 'may not solve the problems of delivering reliable infrastructure to citizens' (p. 31). Attempts at clarifying a viable revenue model for public Wi-Fi, such as Cisco's 'Wi-Fi Opportunity Pyramid' (see TRAI, 2016, p. 12), remain opaque.

What is more, the early scepticism around the economic rationale for municipal (public) Wi-Fi could be said to have been validated by what was occurring to these networks at that time. Looking back on the early 2000s, Laura Forlano has observed that 'municipal networks have struggled to identify appropriate business models, failed to create workable private–public partnerships and, as a result, a number of high-profile projects have been cancelled' (Forlano, 2008, p. 72).

The difficulties early city-based Wi-Fi initiatives encountered in establishing a clear foothold were also not helped by a 'pro-market' stance in certain parts of the world. In the US, for instance, 'incumbent providers and state governments buttressed competition through court and legislative action to restrict municipal investment' (McShane, Wilson, and Meredith, 2014, pp. 128–9). In 2004, Verizon-pushed legislation passed in Pennsylvania which made it more difficult for cities to build wireless networks in that state (Richtel, 2004). A total of fifteen states passed similar anti-municipal broadband laws around that time (Forlano, 2008, p. 82). And, in 2016, the United States Court of Appeals for the Sixth Circuit ruled in favour of the states of North Carolina and Tennessee, overturning a Federal Communications Commission (FCC) order that enabled the towns of Chattanooga, Tennessee, and Wilson, North Carolina, to provide Wi-Fi coverage to residents outside their municipal boundaries (Schwarze, 2018).

A decade further on and the picture looks decidedly different from the portrait we have painted above of the initial struggles municipal Wi-Fi faced. Wi-Fi has since consolidated its position as a vital urban telecommunications and data infrastructure, with multiple – if not always readily discernible or measurable – benefits. As McShane, Wilson, and Meredith explain:

> The commercial, regulatory, technological and social settings that surround public wi-fi have changed substantially in the past few years. The release and rapid uptake of wi-fi-enabled mobile devices, the declining costs and technical complexity

of wireless equipment, moves to engineer seamless network access and handover, and new forms of collaboration between the commercial and public sectors have all underpinned new public investment. (McShane, Wilson, and Meredith, 2014, p. 129)

This sense of optimism around Wi-Fi's present and immediate future is, in part, also attributable to continuous tech development that has been undertaken under the auspices of the Wi-Fi Alliance, including the release of Wi-Fi 6, and the development of other cutting-edge network technologies – such as 5G – that are likely to interact with (rather than supplant) Wi-Fi 6, and a growing awareness of Wi-Fi's larger (if somewhat diffuse) economic and public benefits. And yet some aspects of the social impacts of public Wi-Fi provision, especially around questions of access and equity, remain live issues. It is to these concerns that we now turn in further elaborating on an understanding of public Wi-Fi as a contested space of possibility.

The promise of public Wi-Fi for access and equity

The roll-out of city Wi-Fi raises a number of important questions, including: who has access, and at what cost? Who stands to gain most from its provision? And how well does it service the disadvantaged and the marginalized? In this sense, the provision of city Wi-Fi is firmly situated within ongoing debates about the right to the city (see, for example, Amin and Thrift, 2002, p. 158), with issues of access and equity figuring prominently in the literature on city or municipal Wi-Fi. Within this literature, free public Wi-Fi is regarded as one part of a repertoire of market interventions that hold potential as a 'bridge for the digital divide' (Mackenzie, 2010, p. 54). The term 'digital divide' refers to the gap that exists between those who enjoy ready access to the internet (and associated technologies and infrastructures) and those who do not. As a comparatively low-cost communication technology, Wi-Fi, especially free public Wi-Fi, has been touted as a means of reducing this gap by increasing digital inclusion. The idea of digital inclusion relates to the question of 'whether a person can access, afford and have the digital ability to connect and use online technologies effectively' (Thomas et al., 2019, p. 8). The premise of digital inclusion is that 'everyone should be able to make full use of digital technologies – to manage their health and well-being, access education and services, organise their finances, and connect with friends, family, and the world beyond' (p. 5).

In Chapter 3 of this book, we discussed Wi-Fi as both a literal and figurative gateway: network access is made possible within the home by a 'domestic gateway' (a Wi-Fi router); and the moral economy of the modern technologized home is increasingly defined, negotiated, and defended around, through, and in relation to Wi-Fi. We can extend this same gateway idea to cities and to Wi-Fi provision within the city. Modern cities remain in key respects bounded spaces, with unevenly differentiated forms and levels of access to its spaces and economic and other resources. Wi-Fi provision and access can work to ameliorate as well as reinforce this sense of a city's boundedness and differentiated, uneven access. For instance, one view of urban Wi-Fi, especially municipal free Wi-Fi, is that it might serve as a key gateway that admits into the social and economic life of the modern city people who otherwise do not have access. A competing view is that city Wi-Fi initiatives tend to work as a gateway that principally supports touristic consumers entering the space of the city. Wi-Fi, in this sense, can provide an important mechanism for facilitating, as well as creating barriers to, digital citizenship (McCosker, Vivienne, and Johns, 2016; Isin and Ruppert, 2015; Mossberger, Tolbert, and McNeal, 2007).

Within the scholarship that examines municipal Wi-Fi provision through the lens of these concerns, a key focus has been on mapping geographic diffusion and concentration of Wi-Fi hotspots. What emerges from these studies is a clear sense of an 'unequal spatial distribution of hotspots' (Perng, 2015, p. 187; see also Grubesic and Murray, 2004b). For instance, in a US-based study of Cincinnati Wi-Fi, where all available Wi-Fi hotspots were mapped using wardriving techniques and then analysed, Tony Grubesic and Alan Murray (2004a) found that Wi-Fi hotspots were most heavily clustered in more affluent residential areas, as well as areas associated with education and commerce (such as near to universities and in the CBD), and that Wi-Fi activity was most 'sparse' in more impoverished urban neighbourhoods.

Similar findings emerged from Wang et al.'s (2016) examination of the ambitious 'i-Shanghai' project, involving China Mobile Communications Corporation (CMCC), to roll-out more than 6,000 Wi-Fi hotspots in and around Shanghai. From a spatial analysis of the clustering of these hotspots, Wang et al. note that the project 'deemphasizes issues related to social equity' (p. 11), and conclude that 'the digital divide is still evident especially at the community level' (p. 12). Of these 6,000 hotspots, they observed that only 450 were free (p. 5), and that 'public Wi-Fi access was rarely found in poor residential areas in the inner city and remote rural areas of Shanghai' (p. 12).

In addition to analyses of Wi-Fi clustering, McConnell and Straubhaar (2015) conducted a Bourdieu-inspired study of Wi-Fi access and use within coffee shops in Austin, Texas. While this particular form of Wi-Fi provision represents a relatively low barrier for gaining access to internet services (the price of a cup of coffee), it was rarely utilized by disadvantaged people; rather, these Wi-Fi systems tended to be used by 'members of privileged groups' (p. 225) as a supplement to existing internet use. Low-cost Wi-Fi access alone, they conclude, will not solve digital inclusion issues without also addressing other, wider social barriers to use (p. 228). Even so, this idea of low-cost or free Wi-Fi access has been applied elsewhere within cities, including in McDonald's, supermarkets, malls, and, to cite two Australian examples, hardware stores and certain liquor outlets. It could be that places such as these, with even lower entry costs than cafés and coffee shops, might be the ones that are more frequently accessed and used.

Furthermore, in Australian studies of homelessness in Sydney and Western Sydney, it was found that free Wi-Fi was considered vitally important and was 'the main means for connecting' (Humphry and Pihl, 2016, p. 24; see also Humphry, 2014). Among homeless young people in particular, their tendency to gravitate towards city centres was in part due to the high concentration of Wi-Fi hotspots in these areas (Humphry and Pihl, 2016, p. 24). And yet, while Wi-Fi hotspots abound, they often have conditions associated with their use (such as payment after an initial free trial period) or are concentrated in privately controlled public spaces (such as shopping centres, restaurants, and cafés) (p. 21). The result is that much of the peripatetic movement within the city by homeless youth is undertaken in search of free Wi-Fi. As one study participant puts it, 'I'm walking around and I just have my WiFi open checking trying what crops up on the page, trying to find something that works, usually you can't even find anything anywhere it's pretty hard' (quoted in Humphry and Pihl, 2016, p. 24). Humphry and colleagues advocate for the rolling out of city-wide free Wi-Fi as part of a coordinated response to tackle digital exclusion of low-income and disadvantaged groups (p. 38). (In contrast, two ill-advised and much criticized 'charitable experiments' – one in Prague (Robinson, 2015), and the other set up to coincide with the Austin, Texas-based tech event, South by Southwest (Carmody, 2012) – sought to equip homeless people, in return for receipt of various benefits, with portable routers so that they became roaming Wi-Fi hotspots.) Not addressing the uneven geographical distribution and access issues associated with municipal Wi-Fi that are revealed in the aforementioned studies may well be to risk, as de Waal (2014, p. 143) has warned, taking another step towards a 'splintering urbanism' (Graham and Marvin, 2001) and a deepening digital divide.

Wi-Fi and smart cities

Finally, Wi-Fi can also be seen to operate as a contested space of possibility in relation to smart cities. While Wi-Fi, curiously, doesn't figure much in the academic smart cities literature, in the trade press it is regarded as the 'key' to, or 'backbone' of, smart cities (Fon, 2020; Wi-Fi Attendance, 2018). This is principally because Wi-Fi is a connectivity-enabling or gateway technology; indeed, much of the discussion within the Wi-Fi Alliance-produced literature tends to be framed around the idea of the 'connected' (as opposed to 'smart') city for this very reason. Wi-Fi supports connectivity for the smart city in a number of ways. For instance, Wi-Fi provides 'low cost, easy-to-deploy backhaul for IP [internet protocol] video applications including surveillance, parking management and traffic control' (IDC, 2017). And, while networked things rely on a variety of communication protocols (Bunz and Meikle, 2018, p. 12), Wi-Fi forms an important part of this mix. It continues to play a vital role in enabling urban internet of things and sensor network connectivity (Greengard, 2015, p. 13; Davis-Felner, 2015) due to high adoption of its standards and the relative low cost of its deployment. Wi-Fi is thus regarded as the 'glue that holds smart [connected] cities together' (de Cordova, 2019), and has been described as 'essential fuel' for the 'economic engine' of smart cities (Wi-Fi Attendance, 2018).

Yet the role played by Wi-Fi in supporting smart-city initiatives can be regarded as contested in three ways. First, the benefits that are said to flow from connected cities can only be realized, as Fon's CEO Alex Puregger (2018) points out, if there is continuous, robust city Wi-Fi connectivity. There are, however, questions regarding Wi-Fi's ability to deliver continuous connectivity. To begin with, and as discussed earlier in this chapter, Wi-Fi technology has certain limitations. As Garcia et al. (2018, p. 45) put it, 'WiFi connections [tend to be] less stable than wired ones, the service radius is limited, and it presents high signal attenuation [reduction in the strength of the signal over distance]'. The technology, it has been noted, will need to continue to evolve if Wi-Fi is to continue to play a strategic role in supporting smart cities and IoT (Puregger, 2018). This evolution is occurring, with the development of Wi-Fi HaLow and Wi-Fi 6, both of which are seen as well suited to supporting IoT connectivity (Parekh, 2017), and with the emergence of low-power wide-area network (LPWAN) technologies, such as SigFox, LoRa, NB-IoT, and LTE-M, which allow for the interconnection of devices that only require low-bandwidth connectivity for wireless wide area networks (WBA, 2017). The Wireless Broadband Alliance regards LPWAN technologies as 'promising' in supporting machine-to-machine (M2M) applications, but points out that 'it's still Wi-Fi that carries most of the IoT and smart cities traffic worldwide' (WBA, 2017, p. 34). Wi-Fi service provision will also need to evolve in order to realize robust networks that deliver continuous connectivity. This has led to the rise of Wi-Fi as a service (WaaS) (de Cordova, 2019; GlobeNewswire, 2019), with specialist company partnerships – like that between KodaCloud and Extreme Networks (Businesswire, 2018) – offering cloud-based enterprise Wi-Fi. There has also been significant private-sector investment – by the likes of Cisco, IBM, Intel, and Samsung (Mattern, 2013) – as well as increased public–private partnerships, such as that between the City of New York and Google in the development of LinkNYC free public Wi-Fi.

A second way in which the role played by Wi-Fi in supporting smart-city initiatives can be regarded as contested relates to the roll-out of 5G services. It is worth remembering that 5G is not a standard upgrade to infrastructure and mobile networks (like 3G and 4G). Not only will it replace existing 3G and 4G systems, but also it will vastly expand network capacity, so that cars, utility grids, appliances, medical devices, industrial machinery, cities, and more can all be connected through wide-area networks (Rosenberg, 2020). Telecommunications providers can also 'slice' their network, allowing them to 'provide portions of their networks for specific customer use cases' such as autonomous vehicles or automated machines supporting smart agriculture (Kavanagh, 2018). In this sense, 5G is an emergent technology that is in key respects designed to cater explicitly to the unique demands of 'smart city' initiatives and associated applications. This is not to say, however, that Wi-Fi will no longer continue to play a vital role in these developments.

As noted in Chapter 3, one specific challenge for 5G networks will be their ability to maintain high transmission rates inside buildings (Apostolopoulos, 2019), something that Wi-Fi has been able to achieve. There are a suite of further challenges and issues. There are the high costs associated with building blanket 5G infrastructure throughout cities, the sunk costs in present mobile infrastructure, and existing infrastructural investment in Wi-Fi. On the latter, the Wi-Fi Alliance estimates that there are more than 13 billion Wi-Fi devices in active use worldwide, and many consumer electronics goods and connected devices already come Wi-Fi-enabled. To suddenly switch these off in favour of a 5G network would render these devices defunct (IntechnologySmartCities, n.d.), and arguably opens up holes in and brings instability to the overall 'patchwork quilt' of wired and wireless urban information and communication infrastructures. What is more, there is still significant work to be done around addressing problems with the orchestration of interaction between connected devices that occur as a result of 5G network slicing (Rabie, 2019), and even the orchestration of connection between 5G and Wi-Fi networks. And, at the consumer end, one needs a data plan in order to access 5G, or be tethered to one or to a cellular network, unlike with Wi-Fi. What is more likely is a symbiotic relationship between these two sets of technologies – something that is discussed in more detail in the final chapter. Foreseeing this need for 5G–Wi-Fi complementarity, Australian telecommunications provider Telstra released a 5G / Wi-Fi 6 dongle, the Telstra 5G Wi-Fi Pro (Telstra, 2020). What the release of this device reveals is that 5G doesn't change people's interest in or desire to share connectivity. Consumers tend to want to maximize cost-savings and convenience, regardless of whether they are moving about their homes or their cities.

A third way that Wi-Fi is contested relates to how it becomes enrolled in critiques of smart-city agendas and their impacts. These critiques tend to be focused around a now established set of concerns, which include: what Kitchin (2014, p. 2) terms the 'underlying neoliberal ethos that prioritises market-led and technological solutions to city governance and development'; sociotechnical imaginaries of smart cities (Luque-Ayala and Marvin, 2015; Söderström, Paasche, and Klauser, 2014), which tend to be 'premised in a particular narrative about urban crises and technological salvation' (Sadowski and Bendor, 2019, p. 540); data generation, accumulation, extraction, and capitalization (Sadowski, 2019; Kitchin, 2014, p. 4); and, increased automated surveillance and 'dataveillance' accelerating a new 'spectrum of control' (Sadowski and Pasquale, 2015; Vanolo, 2014). In addition to the above is Sarah Barns's (2020) particularly insightful critique of the 'platformization' of urban life, whereby 'digital platforms have not only become an everyday part of our experience of the city ... they are also seeking to reshape the built fabric' itself (pp. 14–15), generating forms of 'platform urbanism'. What Barns is proposing here also represents a significant step from the attempts at 'platformizing' Wi-Fi (documented at various points throughout this chapter and to be returned to in the final chapter), to Wi-Fi playing a major role in the still developing 'platformization' of cities themselves.

Conclusion

In this chapter, we set out to do three things. We sought to place Wi-Fi (and its infrastructures) within a longer history of city development – a history in which communication technologies take centre stage. We then gave close consideration to two developments that, in combination, worked to accelerate the development and normalization of Wi-Fi use in urban contexts: the rise of Wi-Fi-enabled portable devices, such as laptops and smartphones; and our embrace of cafés as key sites for the provision of Wi-Fi, and as historically significant sites for fostering sociability and communication. Then, we approached city or municipal Wi-Fi as a *contested space of possibility*, employing this as a productive frame or lens through which to view the politics and economics of free public Wi-Fi, and the promise of municipal Wi-Fi – especially in catering to the socially disadvantaged, and in thinking about the role of Wi-Fi in smart cities. At various points throughout the chapter – forming something of a leitmotif – we also gave fleeting consideration to attempts (successful and otherwise) to 'platformize' Wi-Fi, as well as efforts to incorporate it into other, larger, platform efforts and platform ecosystems. What is remarkable about Wi-Fi from this wide-ranging investigation is that, two decades on, this familiar, often-neglected wireless technology has not diminished in its importance. Could it be that Wi-Fi's continued significance in urban contexts, as elsewhere, is, in part, a product of its mutability (Mackenzie, 2010, p. 35), malleability, and functional flexibility?

6

Problems, Prospects, Possibilities

We began this book with the catastrophe of Covid-19. The global pandemic closed down cities, schools, and businesses, impelling people to work from home. In that setting, Wi-Fi became an essential medium for schooling, work, and social connection. However, the benefits of connection were not evenly distributed: while private networks sustained bubbles of safety, people who relied on public Wi-Fi networks in libraries or cafés for internet struggled to retain access. This recent experience underlines the continuing significance of Wi-Fi as both a private and public resource. We argued that, if the pandemic leaves us in a poorer and more unequal world, Wi-Fi in all kinds of settings – not only households – will become increasingly important, because of its capacity to bring affordable, mobile, and accessible internet to more people. In a post-Covid world, different kinds of internet are possible. A more inclusive internet is conceivable, as is a more socially stratified and geographically segmented internet. The long-run reality depends in part on what we do with Wi-Fi.

In this closing chapter, we consider some of Wi-Fi's possible, probable, and preferable futures. Wi-Fi has evolved rapidly and substantially since its early emergence at the turn of the millennium. At a technical level, it is now much better designed for the complexities of internal domestic spaces and larger public installations. It is considerably faster, and somewhat more secure. Early Wi-Fi more or less managed email, printing, file sharing, and the web. Current Wi-Fi streams television, interactive games, online lessons, and audiovisual meetings. Current Wi-Fi has decisively moved beyond the laptop, and more recently beyond the smartphone. A growing community of household and personal devices rely on it, and, as a complement to cellular networks, Wi-Fi occupies a crucial place in phone ecosystems – a position which has given Wi-Fi extraordinary economies of scale.

On the basis of its technical strengths, flexibility, low costs, and deep support across the tech industry, Wi-Fi now has many billions of users. Such a malleable, widely distributed, low-cost communications infrastructure means that future Wi-Fi has extraordinary possibilities. Its prospects and possible uses remain open questions, shaped by the shifting contexts in which Wi-Fi is experienced and understood. As Wi-Fi has evolved, so has the information ecology in which it works, in part because of Wi-Fi's own transformative consequences. Wi-Fi, therefore, is now adapting to an environment it has helped to make. In particular, Wi-Fi has fuelled the emergence of the mobile internet, with both its many benefits and the institutional, economic, and political challenges posed by data-driven digital platforms and the gig economy. The successes of mobile broadband are also double-edged for the proponents of Wi-Fi. They present many specific difficulties: co-existence and competition with cellular networks controlled by telecommunications companies; congestion in the available spectrum; and pressure on the performance of Wi-Fi networks, especially in relation to their security and privacy. There are solutions of a kind in sight for some, but not all, of these problems.

The Wi-Fi Alliance offers its own vision for the future: connecting 'everyone and everything, everywhere'. From healthcare to agriculture, the Alliance is entirely upbeat about the purposes such connections might serve. Hyperbolic and blandly utopian, in the familiar tech mode? Yes, but it's worth unpacking some of the implicit messages in that slogan, which subtly supports a case for open standards and the resources Wi-Fi needs. 'Everyone' and 'everywhere', for example, in this context recognizes two aspirations: the role of Wi-Fi in making the internet more accessible and affordable; and the importance of resisting the geographical segmentation of the internet. Scepticism is also necessary but not sufficient in relation to the 'everything'. Wi-Fi is not necessarily the solution for every kind of connected device: most drones, for example, do use lower-frequency 2.4 GHz Wi-Fi to broadcast video, but, depending on range and terrain, this is not always the best way to communicate with a drone. Hyperbole aside, the successes of Wi-Fi and the network effects already in place mean that it is very likely that the density and coverage of Wi-Fi networks will continue to expand.

In that light, we can acknowledge how important it is for influential organizations such as the Alliance to raise the question of Wi-Fi's possibilities. We can then consider what the Alliance's slogan does not say. What kinds of things could future Wi-Fi actually do, and what problems – for citizens, technologists, policymakers, and students of digital communication – might be submerged in the sales pitch? To address that question, we first consider Wi-Fi's immediate prospects, and what these disclose about the particular dynamics that drive Wi-Fi's transformations. We then turn to possible longer-run futures, before finally considering how Wi-Fi might assist communities and institutions in responding to the pandemic, as well as households.

Wi-Fi in the short run

The immediate future of Wi-Fi is well known and widely publicized. Wi-Fi 6, the Alliance's new 'generational' brand for the IEEE's 802.11ax standard, is now being built into new devices. As with previous new iterations of Wi-Fi, Wi-Fi 6 claims a big step up in speed, with the proviso that actual experience is unlikely to change much for most users. Wi-Fi speeds already comfortably exceed those of most household and small business internet service plans, so faster Wi-Fi makes little difference to download speeds in domestic settings or the local café. The problems with many Wi-Fi networks at home and elsewhere are usually less to do with speed and more to do with the number of devices connected, and the array of different things they may be trying to do. Wi-Fi routers have to manage things that need to happen in real time, such as an online meeting, as well as things which may not be time critical, such as syncing photos. Some devices such as sensors transmit small quantities of data over an extended period, while other devices such as smart TVs want to receive larger streams more occasionally. More bandwidth helps because it can be shared across more connections, so networks should work somewhat faster for everyone. The most important aspects of Wi-Fi 6 are not increased speed, but a few clever strategies (some borrowed from the cellular tech sector) that will help Wi-Fi base stations manage more simultaneous connections for the increasing number and variety of devices. Wi-Fi 6 also includes stronger encryption, and the capacity for scheduled communications, which should reduce battery consumption for connecting devices.

So the objective of Wi-Fi 6 is a faster, more secure network, which can manage a growing number and diversity of connections. There are clues here as to where the Wi-Fi industry sees growth. Institutions such as hospitals are perceived to be one opportunity, where data flows are growing rapidly in importance and complexity for the management of healthcare, with a secondary demand for internet access as a patient and visitor amenity. Another important area for Wi-Fi may be public environments such as stadiums, where spectators can access event-related information and video (and, of course, advertising) alongside the live action. Motor vehicles (especially autonomous vehicles, and including vehicle-to-vehicle communication) are a further likely domain for future Wi-Fi. Some aspects of these possible uses become clearer when we look at Wi-Fi 6 alongside other new Wi-Fi Alliance standards. The Alliance's certified Wi-Fi Data Elements is open source code that aims to formalize the measurement and evaluation of Wi-Fi network performance. That data will be important for capacity trading, advertising, and recommendation systems, among other applications. In this context, it is increasingly clear that Wi-Fi is being reconceived for a growing density of connections, in a wider range of indoor and outdoor settings, supported by a wider range of business models, and with a rapidly expanding range of devices, from medical devices to vehicles (Hetting, 2019; Roberts and Wong, 2019).

Wi-Fi 6 offers a glimpse of how the tech industry sees the evolution of Wi-Fi. It also illustrates the way in which the transformation of Wi-Fi proceeds. In particular, we can see in the case of Wi-Fi 6 something we've already observed, a process we can call 'Wi-Fi gradualism': slow but uneven development of the protocols, a protracted roll-out, a long process of cumulative improvement, and extended periods between major upgrades. (There is an obvious contrast between the gradual evolution of Wi-Fi, and the plasticity of its uses.) While Wi-Fi gradualism has not necessarily been detrimental, it has meant that in some areas, such as privacy and security, Wi-Fi has not kept up with best industry practice. Comparisons are odious, but if we think of the pace of change in technological services or products that have made far-reaching impacts with enormous numbers of users, Wi-Fi has seen six versions in two decades, while twelve major versions of the iPhone (and many more models) have appeared between 2007 and 2020, and Google Search is constantly updated. There is no reason why Wi-Fi should be updated as rapidly as the iPhone, but the differences highlight Wi-Fi's distinctive evolutionary dynamic.

One fundamental aspect of Wi-Fi is that it has always relied on dedicated hardware, a chip (or a group of chips) which manages radio reception and transmission in the designated frequencies, at the required power level, and according to the standard protocols. Wi-Fi circuits are often supplied by specialized suppliers such as the US-based firm Broadcom. Because these chips are made in large numbers and are in the market for many years, they can be very cheap. A Wi-Fi chip that uses the older Wi-Fi standards, of the kind that might be built into a simple 'internet of things' device such as a home lighting system, is likely to cost only a few dollars. (That figure may be compared with the $US99 retail price for the 1999 iBook's optional Wi-Fi card.) However, chips incorporating the most recent standards are much more expensive, and for that reason these are likely to appear first in higher-value devices where Wi-Fi is intensively used, such as laptop computers and smartphones. In the case of smartphones, dedicated processors such as those made by Qualcomm often incorporate Wi-Fi capabilities together with cellular modems, but the need to integrate these capabilities in very resource-constrained devices can further complicate the design and development process.

Wi-Fi's dependence on hardware means that the take-up of new versions of Wi-Fi involves many dependencies. Industry consensus is needed on new or amended standards, but this is always a slow process, and one not required for the iPhone or a search engine. Nevertheless, open standards have the great benefit of reducing the risk of costly fragmentation. They are vital to encourage chip makers to invest in engineering new chipsets. Hardware manufacturers then need an incentive to buy those chips and build them into devices. The bundling of Wi-Fi into the cost of new devices then assists consumer take-up, but consumers are unlikely to buy new devices solely to upgrade their Wi-Fi. Further, in order to benefit from their new Wi-Fi circuitry, consumers must choose to buy updated Wi-Fi gateways or routers, unless these in turn are offered as part of an internet access service. In these circumstances, when the benefits of new Wi-Fi generations are incremental improvements and not all immediately apparent, building a critical consumer base for the new system may take many years, and this in turn will stretch out the upgrade cycle.

However, it is always a mistake to interpret the development of Wi-Fi solely in technical or commercial terms: as we recall, a critical generative moment for early Wi-Fi was the public policy decision to release spectrum for unlicensed use in the United States. Throughout this book, we have emphasized the plasticity of wireless networks, but one major material constraint on their use has always been the amount of radiofrequency spectrum allocated to them, and the nature of that spectrum. Current Wi-Fi works in two blocks of the spectrum, 2.4 GHz and 5 GHz, and as Wi-Fi has grown in popularity, those bands have become steadily more congested. A typical scenario is an apartment block with many closely adjoining Wi-Fi networks, all sharing the same frequencies for an increasing volume of work, entertainment, education, and social connection. The problem manifests not only in slower traffic, but also in users having difficulties connecting to a network even when it is visible. To make matters more complicated, Wi-Fi naturally has no exclusivity over the use of unlicensed bands. For example, the 5 GHz band is already shared with telecommunications companies providing the 4G mobile services known as LTE. These companies use the unlicensed bands in addition to the spectrum they have purchased at auction, in order to manage additional demand in densely busy areas. (Telecommunications companies also offload an increasing amount of cellular traffic onto Wi-Fi as their cellular networks become congested, or in areas where cellular coverage is poor.)

In the light of this spectrum constraint, the most important recent event for the future of Wi-Fi has not been the appearance of Wi-Fi 6, but the decision of the US Federal Communications Commission (FCC) in April 2020 to release a substantial new block of spectrum for unlicensed use, in the 6 GHz band. This is the first significant addition to unlicensed spectrum since the FCC's original allocation in 1989. If the technical and economic dynamic of Wi-Fi is gradualist, the public policy dimension of change moves in much more substantial steps but over a considerably more protracted duration. The allocation of the 6 GHz band gives

Wi-Fi routers around four times the bandwidth they currently work with. More channels will be available, and each channel will be able to carry more information than is possible now with Wi-Fi's current frequencies. The higher 6 GHz frequencies also behave differently from current Wi-Fi's 2.4 GHz band: they allow faster transmission but travel shorter distances.

The new unlicensed US spectrum is expected to stimulate debates and decisions in other jurisdictions. A number of issues are in play, especially the interactions between Wi-Fi and 5G services, but a wide range of existing spectrum users and many other technology and telecommunications companies have interests in the 6 GHz band. In particular, these frequencies are seen as suitable for licensed 5G cellular services. Any technical decisions will inevitably involve political and industry considerations, including proposals to divide spectrum between licensed and unlicensed allocations. Changes in global politics, both within and beyond the tech sectors, will also be important. China has clearly emerged as a leading internet economy in the period since the 1989 FCC decision; more recently, 5G has become a focus for national rivalries and arguments over sovereign security. In China and in international forums, a strong case may be made for allocating further spectrum to licensed 5G, an area where China is considered to have had a technical head start over US and European competitors. The risk is that the debate over 6 GHz and the future of Wi-Fi may be prejudiced by arguments over 5G.

Nevertheless, the FCC decision has created great expectations about stimulating further innovation, research, and development in wireless networks, heavily influenced by the history of early Wi-Fi. In terms of the outcomes for wireless users, there is now an additional element at work: a boost for Wi-Fi, but also a shift which disrupts the Alliance's marketing model of gradual, generational evolution. The Wi-Fi Alliance has branded the new expanded-bandwidth Wi-Fi as Wi-Fi 6E, pairing it with the generational switch to Wi-Fi 6 because it will incorporate the Wi-Fi 6 improvements. But Wi-Fi 6 and 6E will be different steps comprising changes of different kinds, and together they will quicken the pace of change. Regular Wi-Fi 6 equipment will not be able to take advantage of the new spectrum. As we've seen before, the Wi-Fi 6E capabilities will require users to buy another collection of new devices and new routers.

Wi-Fi in the long run

In the short term, the Wi-Fi industry is now well positioned to take advantage of the new unlicensed spectrum and its recent tech revamp. The circumstances of the 2020s are not those of 1989: Wi-Fi is now based on mature technologies which are closely integrated into operating systems and critical internet protocols. The key institutions for standards setting and promoting Wi-Fi networks are well established, and Wi-Fi is, of course, deeply embedded in everyday computing. However, the proponents of Wi-Fi are also by no means the only actors with a stake in how the frequencies are used. A new landscape of corporate and global competition may well emerge in the longer term. Although the FCC refers to the 'unlicensed community', evoking a co-operative domain of experimental technologists, it seems probable that substantial commercial organizations – many of them not yet on the scene in 1989, and all with deep pockets for research and development – will play a major role in determining the uses of this tranche of newly released spectrum. Of course, we should not misconceive or romanticize the commercial origins of a technology designed to connect cash registers, or the scale of the businesses that did most to develop Wi-Fi. NCR in the late 1980s was a major computing company; AT&T, which acquired NCR shortly after the first release of WaveLAN, was a dominant telco. (AT&T opposed the new unlicensed allocation.)

Cellular broadband was a non-existent industry in the era of early Wi-Fi; it is now hugely successful, highly competitive with fixed broadband services, and deeply entrenched in the internet industries. Like Wi-Fi, the cellular industry is in the process of a generational upgrade – marketed as 5G – which offers performance improvements and internet of things capabilities. It is expected that 5G providers will move quickly into any new unlicensed spectrum. They are likely to rely increasingly on Wi-Fi to offload traffic from their own spectrum. Cisco researchers (2018) predict that in the US almost 76 per cent of all mobile data traffic will be offloaded to Wi-Fi by 2022. Most of the large platforms which now dominate the digital economy were also scarcely emergent in the era of early Wi-Fi. A resurgent Apple had the wherewithal to champion Wi-Fi in 1999, but at that point Amazon was still known as a book store, Microsoft was yet to venture into the cloud, Google was but a year old, Alibaba was brand new, and Facebook did not exist. All these firms now have deep interests in internet access, together with the resources and expertise to build their own infrastructures if necessary. We've noted in this book how Wi-Fi has so far resisted incorporation into platform economies: as the stakes are raised, in the longer run, that may become more difficult.

Why might that be so, and what is at stake in Wi-Fi's longer-term future? In this book, we have shown how the adaptable infrastructures of Wi-Fi can positively reconfigure access to the internet in households, communities, and cities. Wi-Fi shares resources and distributes agency in many different kinds of places. It does so using open standards and cheap hardware. It holds out the promise – not always fulfilled – of a more affordable, mobile, and accessible internet. That remains a preferable future, but is it also probable? We have also seen in this book how, over the period of Wi-Fi's emergence, the environments in which we find Wi-Fi are changing: how households, communities, and cities have become increasingly populated with devices, from laptops to phones to doorbell cameras. Wi-Fi has made those transformations possible, but they have led us into a new landscape. Here, Wi-Fi is not merely an 'access technology', a means to share in digital resources through the internet. Instead, it seems probable that we will need to begin to think of Wi-Fi as an infrastructure for generating as well as transmitting data, for activating and linking devices as well as connecting them. The new Wi-Fi standards – not only those intended to support the internet of things but also those concerned with data analytics – facilitate this shift.

Wi-Fi in this scenario is no longer merely enabling digital communication; it is also a means for automation. Further, the kind of automation involved is not merely the simple labour-saving 'smart home' idea of the automated coffee machine or the lights which adjust themselves. It might also include smart speakers which reinforce conservative gender stereotypes (Strengers and Kennedy, 2020), or facial recognition systems which exercise other forms of discrimination. If we think of Wi-Fi's great social and economic attribute as a potential to redistribute agency to people and communities, it may also follow that Wi-Fi will redistribute agency to machines which have the capacity to make decisions for us. Where that does occur, careful design and planning will be necessary to ensure that those data flows and decisions align with human and public interests. It is not beyond us to conceive of ways in which Wi-Fi networks, public and private, could be used to advance human welfare and mitigate the risks of more centralized models of data collection and control.

The sociologist John Urry (2016) wrote about how we might think in social terms about the future, questioning purely technological or commercial probabilities, and considering the social impact of design or policy decisions before they are made. We could think about Wi-Fi's prospects in that way as well, recognizing the many different actors that must be involved, including governments and public authorities, international organizations as well as civil society. Inclusion in this context must extend to participation in decision-making. Consensus in the fashion of the IEEE will never be reached, but it is critically important that public debate around the future of Wi-Fi is supported, and that voices beyond those of technologists and the corporate sector are heard.

Wi-Fi's social futures?

What would an example of 'social futures' thinking about Wi-Fi look like, in the light of the automation scenario we have traced here? We can return to the catastrophes with which we started. We know that home Wi-Fi was vital for those with the resources to use it during the pandemic. The bushfire response, involving the creation of community Wi-Fi facilities when other infrastructures were destroyed (see Figure 1.1), is just as instructive. If we think of Wi-Fi as a uniquely inclusive, decentralized, automated network, generating data as well as merely transmitting it, we can begin to explore ways in which Wi-Fi might help institutions or communities deal with a pandemic or other crisis at a public or local level, beyond simply supporting those able to shelter at home. A simple example demonstrates a possible scenario. In a recent article, University of Melbourne researchers (Dethlefs et al., 2020) showed how Wi-Fi could help to control infection on university campuses, and thereby sustain the use of public spaces.

Some brief background to the problem: from an early point, Covid-19 presented enormous challenges for public health, in part because its dangers were widely underestimated. Testing facilities struggled to cope with the number of tests required, and large numbers of people who should have been tested were not. Those who were tested were not always isolated while waiting for results. Infected people often did not display symptoms of the disease. At one level, these problems can be understood as problems of managing communication across complex social networks, and in some jurisdictions those problems made lockdowns necessary, including the closure of campuses.

In order to track and control the movement of infection through the community, we need to detect those at risk through physical proximity, and then convey the right information to all those who may be at risk. Faced with these problems, governments looked to automated systems for technological solutions. They looked especially to smartphones, which already had the capacity to track locations and identify proximate devices, and therefore (the theory went) the people who used them. If a contact tracing app was downloaded by a sufficiently high percentage of the population, the data it produced could augment more traditional methods for tracing and tracking. In particular, it could help to identify people at risk who were not known to those with the disease. Moreover, such apps could fulfil this function at relatively low cost, because they would rely on existing infrastructure: the network of smartphones and the software distribution platforms created for them.

Governments and other developers chose contact tracing apps using Bluetooth, a short-range wireless standard widely used in smartphones to find and 'pair' nearby devices like headphones, keyboards, or car stereos. Bluetooth devices make connections by broadcasting their availability in frequent short messages, so the theory with automated contact tracing was that, if a Bluetooth message of that kind appeared on a phone from someone found to be infected, then the recipient and anyone with the same message (and the app) could be identified and tested. A strength-of-signal indicator allows the app to estimate the physical distance between smartphones, and therefore people.

In practice, Bluetooth tracing apps have encountered numerous problems. The estimates of physical distance generated through Bluetooth turn out not to be very reliable. But reaching a critical mass of users has been a serious problem. Perhaps 70 per cent of the smartphone user population needs to download a Bluetooth tracing app to make it effective in any given jurisdiction. But, in many countries, uptake has fallen short of that figure. At the time of writing in Australia, the user base for the official COVIDSafe app appeared to be substantially below the target, and the government had stopped actively promoting it. At that point, the app was the eleventh most popular in the free health and fitness category on Apple's app store, ranking just below apps offering daily motivational quotations, and calming stories for meditation.

Without effective tracking and tracing, universities cannot open and may not be able to survive, so the pandemic raises existential problems for such intitutions. Dethlefs and his colleagues point out that existing Wi-Fi networks offer interesting advantages and capabilities compared to Bluetooth. Wi-Fi has a series of distinctive attributes: the campus Wi-Fi network is already used by almost everyone; it covers buildings and open spaces; people log on to the network when they enter campus and log off when they leave; users are always registered by the university network; and Wi-Fi access points indicate where connected users are across campus.

By tracing the movement and density of Wi-Fi users across campus, together with the time they spend in particular locations, universities in this example could identify places of risk where too many people may be present and social distancing will be compromised. These are locations where campus managers can then reconfigure space to encourage distancing, undertake deep cleaning, provide sanitizers, ensure ventilation, and take other steps to reduce the risks of infection. In the event that a person enters the campus who is subsequently identified as contagious, the university can trace the person's progress across the campus, and identify and notify others who may have been at risk. The analysis of people's movements and densities can be done entirely anonymously. Identification of individuals need only occur when a positive case is identified, and that process can be completely separated from the algorithmic analysis.

Dethlefs and his colleagues acknowledge that the example of Wi-Fi contact tracing would involve many additional complexities, not least to do with the security of the networks involved and the data collected, and the consent and agency of the university population. It is interesting, however, that what they propose is not a radical recasting of the uses of Wi-Fi: after all, wireless networks are already used to help to control other aspects of campus management, such as lighting, cooling, and heating.

Wi-Fi was once the key to accessing the internet. Now more than that, it is becoming a pathway to automation. Together with a host of other technical systems, it is taking us in a direction that offers great potential benefits but also carries significant risks. If the social futures of Wi-Fi can be imagined in terms that are true to the creative, adaptive, and inclusive elements of Wi-Fi, the positive possibilities will have a greater chance.

Bibliography

4DI. (2003, 27 August). How Our Unamped Omnis Beat the Vivato™ Switches [Presentation to NYCwireless]. Available at: http://4di.cuzuco.com/2003/4DI.pdf.

ABC News. (2015, 8 April). Free Wi-Fi in Town Camps Could Solve Anti-social Youth Problem in Alice Springs, says CAYLUS. Available at: www.abc.net.au/news/2015-04-08/free-wifi-in-alice-springs-town-camps-could-solve-youth-problems/6378914.

Actiontec. (2020). The Evolution of WiFi Standards: A Look at 802.11a/b/g/n/ac. Available at: www.actiontec.com/wifihelp/evolution-wi-fi-standards-look-802-11abgnac.

Adhikari, R. (2015, 28 January). Tech-Savvy Cubans Build Their Own Private Internet. *TechNewsWorld*. Available at: www.technewsworld.com/story/81646.html.

Aguilar, M. (2014, 18 November). The Plan to Turn NYC's Old Payphones into Free Gigabit Wi-Fit Hot Spots. *Gizmodo Australia*. Available at: www.gizmodo.com.au/2014/11/the-plan-to-turn-old-payphones-into-free-gigabit-wi-fi-hot-spots.

Aldini, A. and Bogliolo, A., eds. (2014). *User-centric Networking: Future Perspectives*. Cham, Switzerland: Springer.

Aldrich, F. K. (2003). Smart Homes: Past, Present, and Future. In R. Harper, ed. *Inside the Smart Home*. London: Springer-Verlag, pp. 17–39.

Alexander, D. T., Barraket, J., Lewis, J. M., and Considine, M. (2012). Civic Engagement and Associationalism: The Impact of Group Membership Scope versus Intensity of Participation. *European Sociological Review*, 28(1), pp. 43–58.

Alliance for Affordable Internet. (2019). *The 2019 Affordability Report*. Washington, DC: Web Foundation.

Amin, A. and Thrift, N. (2002). *Cities: Reimagining the Urban*. Cambridge: Polity.

Anayo, J. and Horst, H. A. (2016). Technologies of the Nation: Public Wi-Fi and the Demand for More in Niue. *Information Technologies and International Development*, 12(4), pp. 1–9.

Anderson, C. (2003, 1 May). The Wi-Fi Revolution. *Wired*. Available at: www.wired.com/2003/05/wifirevolution.

Andrejevic, M. (2007). Surveillance in the Digital Enclosure. *The Communication Review*, 10, pp. 295–317.

Andrews, E. L. (1994, 28 April). Bell Atlantic Set to Offer Wireless Data Service. *New York Times*. Available at: https://global-factiva-com.

APC-IDRC. (2018). *Global Information Society Watch 2018: Community Networks*. Johannesburg, South Africa, and Ottawa, Canada: Association of Progressive Communications (APC) / International Development Research Centre (IDRC).

Apostolopoulos, J. (2019, 9 April). Why Wi-Fi 6 and 5G Are Different: Physics, Economics, and Human Behavior [blog]. Cisco. Available at: https://blogs.cisco.com/wireless/why-wifi6-and-5g-are-different.

Arar, Y. (2003, 4 December). Pocket PCs: The Wi-Fi Generation. *PCWorld*. Available at: www.pcworld.com/article/113650/article.html.

Armitage, G., Kennedy, J., Nguyen, S., Thomas, J., and Ewing, S. (2017, January). *Household Internet and the 'Need for Speed': Evaluating the Impact of Increasingly Online Lifestyles and the Internet of Things*. Centre for Advanced Internet Architectures, Technical Report 170113A. Melbourne, Australia: Centre for Advanced Internet Architectures, Swinburne University of Technology. Available at: http://caia.swin.edu.au/reports/170113A/CAIA-TR-170113A.pdf.

Arnold, M., Nansen, B., Kennedy, J., Gibbs, M., Harrop, M., and Wilken, R. (2016). An Ontography of Broadband on a Domestic Scale. *Transformations*, 27. Available at: www.transformationsjournal.org/wp-content/uploads/2016/12/Arnold_et_al_Transformations27.pdf.

Arthur, W. B. (2010). *The Nature of Technology: What It Is and How It Evolves*. London: Penguin Books.

Associated Press. (2009, 15 December). McDonald's to Offer Free Wi-Fi. *CBS News*. Available at: www.cbsnews.com/news/mcdonalds-to-offer-free-wifi.

Austin, P. L. (2018, 15 May). The Wi-Fi Alliance Wants to Make Building a Mesh Network Simpler than Ever. *Gizmodo Australia*. Available at: www.gizmodo.com.au/2018/05/the-wi-fi-alliance-wants-to-make-building-a-mesh-network-simpler-than-ever.

Baker, T. (2004, October). Your Data Is on the Air: Connect Your Home and Cut the Wires. *CE Tips*. Available at: https://archive.org/details/ce-tips-magazine-v2i10/page/n37/mode/2up.

Bar, F. and Galperin, H. (2004). Building the Wireless Internet

Infrastructure: From Cordless Ethernet Archipelagos to Wireless Grids. *Communications & Strategies*, 54, pp. 45–68.

Bar, F. and Park, N. (2006). Municipal Wi-Fi Networks: The Goals, Practices, and Policy Implications of the U.S. Case. *Communications & Strategies*, 61, pp. 107–25.

Barns, S. (2020). *Platform Urbanism: Negotiating Platform Ecosystems in Connected Cities*. Singapore: Palgrave Macmillan.

Bateyko, D. (2017, 8 June). Athens's Community WiFi Project Exarcheia Net Brings Internet to Refugee Housing Projects. P2P Foundation. Available at: https://blog.p2pfoundation.net/athens-community-wifi-project-exarcheia-net-brings-internet-refugee-housing-projects/2017/06/08.

Bausinger, H. (1984). Media, Technology and Daily Life. Trans. L. Jaddou and J. Williams. *Media, Culture and Society*, 6, pp. 343–51.

Behrendt, L. (1995). *Aboriginal Dispute Resolution: A Step Towards Self-Determination and Community Autonomy*. Leichhardt, NSW: Federation Press.

Belli, L., ed. (2016). *Community Connectivity: Building the Internet from Scratch. Annual Report of the UN IGF Dynamic Coalition on Community Connectivity*. Rio de Janeiro, Brazil: FGV Direito Rio.

Benkler, Y. (2006). *The Wealth of Networks: How Social Production Transforms Markets and Freedom*. New Haven and London: Yale University Press.

Berker, T., Hartmann, M., Punie, Y., and Ward, K. (2006). Introduction. In T. Berker, M. Hartmann, Y. Punie, and K. Ward, eds. *Domestication of Media and Technology*. Maidenhead, Berks: Open University Press, pp. 1–17.

Berlant, L. (2016). The Commons: Infrastructure for Troubling Times. *Environment and Planning D: Society and Space*, 34(3), pp. 393–419.

Bhagat, A. (2008). Life After Connectivity: The Impact of the Community Mesh Network in Mahavilachchiya, Sri Lanka's E-Village. *The Journal of Community Informatics*, 4(1). Available at: www.ci-journal.net/index.php/ciej/article/view/425/391.

Birdman, J. M. (2016). 'No Wi-Fi, No Snacks, No Friends': Adolescents in Iceland Discussing Media. MA thesis, The University of Iceland.

Blackburn, R. (2017, 22 December). The Secret Life of Hedy Lamarr. *Science*, 358(6370), p. 1546.

Bly, S., Schilit, B., McDonald, D. W., Rosario, B., and Saint-Hilaire, Y. (2006). Broken Expectations in the Digital Home. In *Proceedings of CHI 2006, April 22–24, 2006, Montréal, Québec, Canada*, pp. 568–73. Available at: https://dl.acm.org/doi/pdf/10.1145/1125451.1125571.

Bowker, G. C. and Star, S. L. (1999). *Sorting Things Out: Classification and its Consequences*. Cambridge, MA: MIT Press.

Bravo, J. L. P., Mpfumo, R. B., Milanés, L. A. M., et al. (2018). Lessons from El Paquete, Cuba's Offline Internet. In: *COMPASS '18, June 20–22, 2018, Menlo Park and San Jose, CA, USA*. Available at: https://doi.org/10.1145/3209811.3209876.

Bunz, M. and Meikle, G. (2018). *The Internet of Things*. Cambridge: Polity.

Busch, L. (2011). *Standards: Fecipes for Reality*. Cambridge, MA: MIT Press.

Businesswire, (2018, 5 June). KodaCloud Announces Partnership with Extreme Networks to Bring the Power of Cloud-based Artificial Intelligence to Enterprise WLANs. Available at: www.businesswire.com/news/home/20180605005119/en/KodaCloud-Announces-Partnership-Extreme-Networks-Bring-Power.

Cain Miller, C. (2010, 14 June). Aiming at Rivals, Starbucks Will Offer Free Wi-Fi. *New York Times*. Available at: www.nytimes.com/2010/06/15/technology/15starbux.html.

Carmody, T. (2012, 12 March). The Damning Backstory Behind 'Homeless Hotspots' at SXSW. *Wired*. Available at: www.wired.com/2012/03/the-damning-backstory-behind-homeless-hotspots-at-sxswi.

Carpentier, N. (2008). The Belly of the City: Alternative Communicative City Networks. *The International Communication Gazette*, 70(3–4), pp. 237–55.

Castells, M. (1989). *The Informational City: Information Technology, Economic Restructuring, and the Urban-Regional Process*. Oxford: Basil Blackwell.

CAT. (2016). *The Northern Territory Homelands and Outstations Assets and Access Review. Final Report*. Alice Springs: Centre for Appropriate Technology Ltd (CAT).

Cattoni, R. (2017). *Convergent Technology Research for Application in Remote Indigenous Media Space*. Alice Springs: Rita Cattoni. Available at: www.churchilltrust.com.au/media/fellows/Cattoni_R_2017_Convergent_technology_research_for_application_in_remote_Indigenous_media.pdf.

Chia, P. (2003). The Cantenna Project. *Atomic: Maximum Power Computing*. Available at: https://archive.org/details/Atomic-034-2003-11/page/n101/mode/2up/search/Wi-Fi?q=Wi-Fi.

Chigona, W., Mudavanhu, S. L., Siebritz, A., and Amerika, Z. (2016). Domestication of Free Wi-Fi Amongst People Living in Disadvantaged Communities in Western Cape Province of South Africa. In *SAICSIT '16, 26–28 September, 2016, Johannesburg, South*

Africa. New York: ACM. Available at: https://doi.org/10.1145/2987491.2987500.

Cho, M.-H. (2020, 21 August). KT to Expand South Korea's Free Public Wi-Fi Availability. ZDNet. Available at: www.zdnet.com/article/kt-to-expand-south-koreas-free-public-wi-fi-availability/#ftag=CAD-00-10aag7e.

Cisco. (2016). *Global – 2012 Forecast Highlights*. Available at: www.cisco.com/c/dam/m/en_us/solutions/service-provider/vni-forecast-highlights/pdf/Global_2021_Forecast_Highlights.pdf.

Cisco. (2018). *VNI Mobile Forecast Highlights Tool*. Available at: www.cisco.com/c/m/en_us/solutions/service-provider/forecast-highlights-mobile.html.

Coase, R. H. (1959/2013, November). The Federal Communications Commission. *The Journal of Law & Economics*, 56(4), pp. 879–915.

Coffman, T. (2003). *Building for Hearst and Morgan: Voices from the George Loorz Papers*. Berkeley Hill Books.

Cohen, N. (2014, 5 October). Hong Kong Protests Propel FireChat Phone-to-Phone App. *New York Times*. Available at: www.nytimes.com/2014/10/06/technology/hong-kong-protests-propel-a-phone-to-phone-app-.html.

Coldewey, D. (2013, 16 January). Lost Phones Keep Being Tracked to House in Vegas – to Owner's Frustration. *Today*. Available at: www.today.com/money/lost-phones-keep-being-tracked-house-vegas-owners-frustration-1B8003116.

Communications Technology. (2009). Wi-Fi Rules the Home. *Communications Technology*, 26(11), p. 27.

Condon, S. (2018, 17 September). Shared Spectrum: What's Next and Why It Matters for Businesses. ZDNet. Available at: www.zdnet.com/article/shared-spectrum-whats-next-and-why-it-matters-for-businesses.

Constine, J. (2013, 2 October). Facebook and Cisco Let Brick-and-Mortars Demand Customers Check-in to Get Wi-Fi. TechCrunch. Available at: http://techcrunch.com/2013/10/02/facebookwifi.

Coombe, R. (1998). *The Cultural Life of Intellectual Properties*. Durham, NC: Duke University Press.

Cooper, G. (2002). Mutable Mobiles: Social Theory in a Wireless World. In B. Brown, N. Green, and R. Harper, eds. *Wireless World: Social and Interactional Aspects of the Mobile Age*. London: Springer, pp. 19–31.

Crist, R. (2019, 22 December). *Wi-Fi Is Faster Than Ever: Here's Where it's Headed in 2020*. CNet. Available at: www.cnet.com/news/wi-fi-is-faster-than-ever-heres-where-its-headed-in-2020.

David, P. A. (1985). Clio and the Economics of QWERTY. *The American Economic Review*, 75(2), pp. 332–7.

Davis-Felner, K. (2015, 13 October). Wi-Fi and the Rise of Smart Cities. TMCnet. Available at: www.tmcnet.com/voip/departments/articles/411301-wi-fi-the-rise-smart-cities.htm.

de Camargo Penteado, C. L., de Souza, P. R. E., and Amadeu da Silveira, S. (2017). Connectivity Public Policy in the Network Society: The Case of 'Wifi Livre SP'. *Studies in Media and Communications*, 12, pp. 299–314.

de Cordova, S. F. (2019, 17 May). Wi-Fi as a Service Removes a Major Challenge to Smart City Success. *Medium*. Available at: https://medium.com/smartcityworld/wi-fi-as-a-service-removes-a-major-challenge-to-smart-city-success-331050f3c2eb.

De Filippi, P. and Tréguer, F. (2015). Expanding the Internet Commons: The Subversive Potential of Wireless Community Networks. *Journal of Peer Production*, May. Available at: http://peerproduction.net/wp-content/uploads/2015/01/De-Filippi-Tr%C3%A9guer-Expanding-the-Internet-Commons-with-Community-Networks.pdf.

de Waal, M. (2014). *The City as Interface: How New Media Are Changing the City*. Rotterdam: nao10 publishers.

Department of Communication and the Arts. (2017, May). *Radiocommunications Bill 2017: A Platform for the Future. Information Paper*. Canberra: Australian Government. Available at: www.communications.gov.au.

Dethlefs, J., Wei, S., Winter, S. and Tomko, M. (2020, 6 August). How to Use Wi-Fi Networks to Ensure a Safe Return to Campus. *IEEE Spectrum*. Available at: https://spectrum.ieee.org/view-from-the-valley/telecom/wireless/want-to-return-to-campus-safely-tap-wifi-network.

Dixon, R. C. (1975). Why Spread Spectrum? *IEEE Communications Society Magazine*, 14(4), pp. 21–5.

Dobush, G. (2016, 28 May). Why Is It Impossible to Find Free Wi-Fi in Germany? *Quartz*. Available at: https://qz.com/694618/why-is-it-impossible-to-find-free-wi-fi-in-germany.

Doctorow, C. (2005, 8 November). WiFi Isn't short for 'Wireless Fidelity'. Boing Boing. Available at: https://boingboing.net/2005/11/08/wifi-isnt-short-for.html

Doyle, M. R. (2015). Designing for Mobile Activities: WiFi Hotspots, Users, and the Relational Programming of Place. In A. de Souza e Silva and M. Sheller, eds. *Mobility and Locative Media: Mobile Communication in Hybrid Spaces*. New York: Routledge, pp. 188–206.

Duckett, C. (2019, 4 July). Stop Using 2.4GHz and Rely on 5GHz

Wi-Fi: ACMA NBN Modem Study. ZDNet. Available at: www.zdnet.com/article/stop-using-2-4ghz-and-rely-on-5ghz-wi-fi-acma-nbn-modem-study.

Dunne, A. (2005). *Hertzian Tales: Electronic Products, Aesthetic Experience, and Critical Design*. Cambridge, MA: MIT Press.

Dunne, A. and Raby, F. (2001). *Design Noir: The Secret Life of Electronic Objects*. London and Basel: August Media Ltd / Birkhäuser.

Dutton, W. H. (2005). The Internet and Social Transformation: Reconfiguring Access. In W. H. Dutton, B. Kahin, R. O'Callaghan, and A. W. Wykoff, eds. *Transforming Enterprise*. Cambridge, MA: MIT Press, pp. 375–97.

Echániz, N. and López Pezé, F. (2018). Decentralising Culture: The Challenge of Local Content in Community Networks. In APC-IDRC, ed. *Global Information Society Watch 2018: Community Networks*. Johannesburg, South Africa, and Ottawa, Canada: Association for Progressive Communications (APC) / International Development Research Centre (IDRC), pp. 52–6.

Economist. (2003, 20 December). The Internet in a Cup; Coffeehouses. *The Economist*, pp. 88–90.

Economist. (2004, 12 June). A Brief History of Wi-Fi. *Economist Technology Quarterly*. Available at: www.economist.com/technology-quarterly/2004/06/12/a-brief-history-of-wi-fi.

Edwards, P. N. (2003). Infrastructure and Modernity: Force, Time, and Social Organization in the History of Sociotechnical Systems. In T. J. Misa, P. Brey, and A. Feenberg, eds. *Modernity and Technology*. Cambridge, MA: MIT Press, pp. 185–225.

Edwards, P. N., Jackson, S. J., Bowker, G. C., and Knobel, C. P. (2007, January). *Understanding Infrastructure: Dynamics, Tensions, and Design. A Report of a Workshop on 'History & Theory of Infrastructure: Lessons for New Scientific Cyberinfrastructures'*. Alexandria, VI: National Science Foundation.

Ellis, M. (2004). *The Coffee House: A Cultural History*. London: Weidenfeld and Nicolson.

Estes, A. C. (2015, 27 January). Cuba's Illegal, Porn-Free Underground Internet Is Thriving. Gizmodo Australia. Available at: www.gizmodo.com.au/2015/01/cubas-illegal-underground-internet-is-thriving.

Ethos Global. (2014). *Wi-Fi Rollout CLO Data Collection. Report to the Department of the Prime Minister and Cabinet*. FOI request number: FOI-2017-177.

Falls, R. (2018). SNET: The Shortest Distance. *La Habana*. Available at: www.lahabana.com/content/snet-shortest-distance.

FCC. (1984). Further Notice of Inquiry and Notice of Proposed Rule

Making adopted regarding authorization of spread spectrum and other wideband emissions not presently provided for in FCC Rules and Regulations. Notice of Proposed Rulemaking. General Docket #: FCC-81-413. In *Federal Communications Commission Reports: Advance Reports*. Washington, DC: Federal Communications Commission, pp. 380–401.

FCC. (2003). Facilitating Opportunities for Flexible, Efficient, and Reliable Spectrum Use Employing Cognitive Radio Technologies. Notice of Proposed Rulemaking. DA/FCC #: FCC-03-322. Federal Communications Commission. Available at: www.fcc.gov/document/facilitating-opportunities-flexible-efficient-and-reliable-spectrum-1.

First Mile Connectivity Consortium. (2018). *Stories from the First Mile: Digital Technologies in Remote and Rural Indigenous Communities*. Fredericton, Canada: First Nations Innovation and First Mile Connectivity Consortium.

Fischer, C. S. (1987). The Revolution in Rural Telephony, 1900–1920. *Journal of Social History*, 21(1), pp. 5–26.

Fischer, C. S. (1992). *America Calling: A Social History of the Telephone to 1940*. Berkeley and Los Angeles: University of California Press.

FNMA. (2019). *Communiqué – Indigenous Focus Day 2019 'Shaping Our Digital Futures – Apurte Akaltye-antheme'*. Alice Springs: First Nations Media Australia.

Fon. (2020). Public WiFi Is Key to Smart Cities [blog]. Available at: https://fon.com/blog-public-wifi-smart-cities.

Forlano, L. (2008). When Code Meets Place: Collaboration and Innovation at WiFi Hotspots. PhD thesis, Columbia University.

Forlano, L. (2009). WiFi Geographies: When Code Meets Place. *The Information Society*, 25(5), pp. 344–52.

Fortunati, L. (2018). Robotization and the Domestic Sphere. *New Media & Society*, 20(8), pp. 2673–90.

Forty, A. (2004). *Words and Buildings: A Vocabulary of Modern Architecture*. London: Thames and Hudson.

Frangoudis, P. A., Polyzos, G. C., and Kemerlis, V. P. (2011). Wireless Community Networks: An Alternative Approach for Nomadic Broadband Network Access. *IEEE Communications Magazine*, May, pp. 206–13.

Frosio, G. F. (2018). Why Keep a Dog and Bark Yourself? From Intermediary Liability to Responsibility. *International Journal of Law and Information Technology*, 26, pp. 1–33.

Fuentes-Bautista, M. and Inagaki, N. (2006). Reconfiguring Public Internet Access in Austin, TX: Wi-Fi's Promise and Broadband Divides. *Government Information Quarterly*, 23, pp. 404–34.

Fuller, M. (2007). *Media Ecologies: Materialist Energies in Art and Technoculture*. Cambridge, MA: MIT Press.

Galloway, A. R. (2004). *Protocol: How Control Exists after Decentralization*. Cambridge, MA: MIT Press.

Garcia, L., Jiménez, J. M., Taha, M., and Lloret, J. (2018). Wireless Technologies for IoT in Smart Cities. *Network Protocols and Algorithms*, 1(1), pp. 23–64.

Garton, A. (2018). Australia. In APC-IDRC, ed. *Global Information Society Watch 2018: Community Networks*. Johannesburg, South Africa, and Ottawa, Canada: Association for Progressive Communications (APC) / International Development Research Centre (IDRC), pp. 67–72.

Gates, B. (1996). *The Road Ahead*. New York: Penguin.

Giovanella, F. (2016). Community Networks: Legal Issues, Possible Solutions and a Way Forward in the European Context. In L. Belli, ed. *Community Connectivity: Building the Internet from Scratch. Annual Report of the UN IGF Dynamic Coalition on Community Connectivity*. Rio de Janeiro, Brazil: FGV Direito Rio, pp. 111–22.

GlobeNewswire. (2019, 26 March). Global WiFi as a Service Market Forecast to 2023 – Increased Demand for WiFi as a Service in Small and Medium and Distributed Enterprises. Available at: www.globenewswire.com/news-release/2019/03/26/1768393/0/en/Global-WiFi-as-a-Service-Market-Forecast-to-2023-Increased-Demand-for-WiFi-as-a-Service-in-Small-and-Medium-and-Distributed-Enterprises.html.

Goggin, G. (2006). *Cell Phone Culture*. London: Routledge.

Goggin, G. (2007). An Australian Wireless Commons? *Media International Australia*, 125, pp. 118–30.

Goggin, G. (2011). *Global Mobile Media*. London: Routledge.

Goodman, D. (1994). *The Republic of Letters: A Cultural History of the French Enlightenment*. Ithaca, NY: Cornell University Press.

Graham, S. and Marvin, S. (1996). *Telecommunications and the City: Electronic Spaces, Urban Places*. London: Routledge.

Graham, S. and Marvin, S. (2001). *Splintering Urbanism: Networked Infrastructures, Technological Mobilities and the Urban Condition*. London: Routledge.

Greengard, S. (2015). *The Internet of Things*. Cambridge, MA: MIT Press.

Gross, G. (2018, 25 September). New York City Groups Take Broadband into their Own Hands. Internet Society. Available at: www.internetsociety.org/blog/2018/09/new-york-city-groups-take-broadband-into-their-own-hands.

Grubesic, T. and Murray, A. (2004a). 'Where' Matters: Location and Wi-Fi Access. *Journal of Urban Technology*, 11(1), pp. 1–28.

Grubesic, T. and Murray, A. (2004b). Waiting for Broadband: Local Competition and the Spatial Distribution of Advanced Telecommunication Services in the United States. *Growth & Change*, 35, pp. 139–65.

guifi.net. (2009, 7 June). What is Guifi.net? Available at: https://guifi.net/en/what_is_guifinet.

Gumpert, G. and Drucker, S. J. (2008). Communicative Cities. *International Communication Gazette*, 70(3–4), pp. 195–208.

Guo, E. (2017, 22 December). Without Net Neutrality, Is It Time to Build Your Own Internet? Inverse. Available at: www.inverse.com/article/39507-mesh-networks-net-neutrality-fcc.

Gural, N. (2010, 15 January). McDonald's Serves Up Free Wi-Fi. *Forbes*. Available at: www.forbes.com/2010/01/15/mcdonalds-wifi-starbucks-markets-equities-free-wifi.html#b57d83337f76.

Gurstein, M. (2007). *What Is Community Informatics (and Why Does It Matter)?* Milan: Polimetrica.

Gurstein, M. (2012). Towards a Conceptual Framework for a Community Informatics. In A. Clement, M. Gurstein, and G. Longford, eds. *Connecting Canadians: Investigations in Community Informatics*. Edmonton: Athabasca University Press, pp. 35–60.

Habermas, J. (1989). *The Structural Transformation of the Public Sphere: An Inquiry into a Category of Bourgeois Society*. Trans. T. Burger and F. Lawrence. Cambridge: Polity.

Haddon, L. (1994). Explaining ICT Consumption: The Case of the Home Computer. In R. Silverstone and E. Hirsch, eds. *Consuming Technologies: Media and Information in Domestic Spaces*. London: Routledge, pp. 82–96.

Haddon, L. (2006). The Contribution of Domestication Research to In-Home Computing and Media Consumption. *The Information Society*, 22(4), pp. 195–203. Available at: https://doi.org/10.1080/01972240600791325.

Haddon, L. (2007). Roger Silverstone's Legacies: Domestication. *New Media & Society*, 9(1), pp. 25–32.

Haddon, L. (2011). Domestication Analysis, Objects of Study, and the Centrality of Technologies in Everyday Life. *Canadian Journal of Communication*, 36(2), pp. 311–23.

Haddon, L. (2016). The Domestication of Complex Media Repertoires. In K. Sandvik, A. M. Thorhauge, and B. Valtysson, eds. *The Media and the Mundane: Communication Across Media in Everyday Life*. Göteborg, Sweden: Nordicom, pp. 17–30.

Haine, W. S. (1999). *The World of the Paris Café: Sociability Among the French Working Class, 1789–1914*. Baltimore: Johns Hopkins University Press.

Hajela, S. (2013, 2 October). Facebook and Cisco Connect Businesses and Customers [blog]. Cisco. Available at: http://blogs.cisco.com/news/facebook-and-cisco-connect-businesses-and-consumers.

Hamilton, L. (2017, 10 January). WiFi Is By Far the Most Common Way U.S. Broadband Households Access the Internet: Report. EE World Online. Available at: www.eeworldonline.com/wifi-is-by-far-the-most-common-way-u-s-broadband-households-access-the-internet-report.

Hampton, K. N. and Gupta, N. (2008). Community and Social Interaction in the Wireless City: Wi-Fi Use in Public and Semi-Public Spaces. *New Media & Society*, 10(6), pp. 831–50.

Hampton, K. N., Livio, O. and Sessions Goulet, L. (2010). The Social Life of Wireless Urban Spaces: Internet Use, Social Networks, and the Public Realm. *Journal of Communication*, 60, pp. 701–22.

Hargreaves, T. and Wilson, C. (2017). *Smart Homes and their Users*. Cham, Switzerland: Springer.

Harper, R. (2011). From Smart Home to Connected Home. In R. Harper, ed. *Connected Homes: The Future of Domestic Life*. London: Springer, pp. 3–18.

Hawkins, E. (2017, 29 November). Cable Is for TV. *Electronic Musician*. Available at: www.emusician.com/how-to/cable-is-for-tv.

Hayes, V. and Lemstra, W. (2009). Licence-exempt: The Emergence of Wi-Fi. *Info*, 11(5), pp. 57–71.

Hazlett, T. W., Porter, D. and Smith, V. (2011, November). Radio Spectrum and the Disruptive Clarity of Ronald Coase. *The Journal of Law & Economics*, 54(4), pp. S125–S165.

Healy, T. (2019). Wi-Fi Router. In C. Op den Kamp and D. Hunter, eds. *A History of Intellectual Property in Fifty Objects*. Cambridge University Press, pp. 377–84.

Heer, T., Hummen, R., Viol, N., Wirtz, H., Götz, S., and Wehrle, K. (2010). Collaborative Municipal Wi-Fi Networks – Challenges and Opportunities. In *Proceedings of 8th IEEE International Conference on Pervasive Computing and Communications*. New York: IEEE, pp. 588–93. Available at: https://doi.org/10.1109/PERCOMW.2010.5470505.

Henriksen, I. M. and Tjora, A. (2018). Situational Domestication and the Origin of the Cafe Worker Species. *Sociology*, 52(2), pp. 351–66.

Hesmondhalgh, D. and Lobato, R. (2019). Television Device Ecologies,

Prominence and Datafication: The Neglected Importance of the Set-top Box. *Media, Culture & Society*, 41(7), pp. 958–74.

Hetting, Claus, (2019, 3 July). New Wi-Fi Diagnostics & Data Collection Standard a 'Game Changer', Says ASSIA. *Wi-Fi Now*. Available at: https://wifinowglobal.com/news-and-blog/new-wi-fi-diagnostics-data-collection-standard-a-game-changer-says-assia.

Heyden, T. (2012, 18 October). The Rise of Passive-Aggressive Wi-Fi Names. *BBC News Magazine*. Available at: www.bbc.com/news/magazine-19760006.

Hills, A. (2011). *Wi-Fi and the Bad Boys of Radio: Dawn of a Wireless Technology*. Indianapolis: Dog Ear Publishing.

Hjorth, L., Burgess, J., and Richardson, M., eds. (2012). *Studying Mobile Media: Cultural Technologies, Mobile Communication, and the iPhone*. New York: Routledge.

Hoffman, C. (2018, 21 October). What Is WiGig, and How Is It Different from Wi-Fi 6? How-To Geek. Available at: www.howtogeek.com/371328/what-is-wigig-and-how-is-it-different-from-wi-fi-6.

Hoffman, C. (2019, 6 January). Wi-Fi 6: What's Different, and Why It Matters. How-To Geek. Available at: www.howtogeek.com/368332/wi-fi-6-what%E2%80%99s-different-and-why-it-matters.

Holley, P. (2016, 26 January). People Keep Going to His Home Looking for Their Lost Phones – and Nobody Knows Why. *Chicago Tribune*. Available at: www.chicagotribune.com/ct-lost-phone-tracking-app-leads-to-wrong-house-20160126-story.html.

Horn, C. and Rennie, E. (2018). Digital Access, Choice and Agency in Remote Sarawak. *Telematics and Informatics*, 35, pp. 1935–48.

Horn, C., Rennie, E., Gifford, S., Riman, R. M., and Wee, G. L. H. (2018). *Digital Inclusion and Mobile Media in Remote Sarawak*. Melbourne: Swinburne University of Technology. Available at: https://doi.org/0.4225/50/5a5e751012e2e.

Humphry, J. (2014, August). *Homeless and Connected: Mobile Phones and the Internet in the Lives of Homeless Australians*. Sydney: The University of Sydney / Australian Communications Consumer Action Network (ACCAN). Available at: https://apo.org.au/node/40723.

Humphry, J. and Pihl, K. (2016, June). *Making Connections: Young People, Homelessness and Digital Access in the City*. Abbotsford, Victoria: Young and Well Cooperative Research Centre (CRC). Available at: https://apo.org.au/node/68481.

IDC. (2017). The Role of Public Wi-Fi in Enabling Smart Cities: Business Models and Use Cases for Maximum Impact. *Commscope*. Available at: www.commscope.com/globalassets/digizuite/1065-

1055-role-of-public-wifi.pdf?utm_source=ruckus&utm_medium= redirect.
IDC. (2019, 29 March). Double-Digit Growth Expected in the Smart Home Market, Says IDC. Available at: www.idc.com/getdoc.jsp?containerId=prUS44971219.
IHS Markit. (2008, 29 December). Notebook PC Shipments Exceed Desktops for First Time in Q3. Available at: https://technology.ihs.com/393634/notebook-pc-shipments-exceed-desktops-for-first-time-in-q3.
IntechnologySmartCities. (n.d.). 5G Versus WiFi? The Connectivity Debate [blog]. Available at: www.intechnologywifi.com/blog/5g-versus-wifi-the-connectivity-debate.
International Monetary Fund. (2020). *World Economic Outlook October 2020: A Long and Difficult Ascent*. Washington, DC: International Monetary Fund.
IRCA. (2017). *19th Remote Indigenous Media Festival Magazine*. Alice Springs: Indigenous Remote Communications Association (IRCA).
Isin, E. and Ruppert, E. (2015). *Being Digital Citizens*. New York: Rowman & Littlefield.
Johnson, L. and Lloyd, J. (2004). *Sentenced to Everyday Life: Feminism and the Housewife*. Oxford: Berg.
Jungnickel, K. (2014). *DiY WiFi: Re-imagining Connectivity*. Houndmills, Basingstoke: Palgrave Macmillan.
Kalathil, S. and Boas, T. C. (2001). The Internet and State Control in Authoritarian Regimes: China, Cuba, and the Counterrevolution. *First Monday*, 6(8). Available at: https://firstmonday.org/ojs/index.php/fm/article/view/876/785.
Kanellos, M. (2009, 18 March). PCs: More Than 1 Billion Served. CNet. Available at: www.cnet.com/news/pcs-more-than-1-billion-served.
Kastrenakes, J. (2019, 21 February). Wi-Fi 6: Is it Really That Much Faster? *The Verge*. Available at: www.theverge.com/2019/2/21/18232026/wi-fi-6-speed-explained-router-wifi-how-does-work.
Kavanagh, S. (2018, 28 August). What Is Network Slicing? 5G.co.uk. Available at: https://5g.co.uk/guides/what-is-network-slicing/.
Kelty, C. M. (2005). Geeks, Social Imaginaries, and Recursive Publics. *Cultural Anthropology*, 20(2), pp. 185–214.
Kelty, C. M. (2008). *Two Bits: The Cultural Significance of Free Software*. Durham, NC: Duke University Press.
Kennedy, J., Arnold, M., Gibbs, M., Nansen, B., and Wilken, R. (2020). *Digital Domesticity: Media, Materiality, and Home Life*. New York: Oxford University Press.

Kiesler Markey, H. and Antheil, G. (1941). Secret Communication System. Patent US2292387A. Available at: https://patents.google.com/patent/US2292387A/en.

Kitchin, R. (2014). The Real-time City? Big Data and Smart Urbanism. *GeoJournal*, 79, pp. 1–14.

Kleinman, S. (2006). Cafe Culture in France and the United States: A Comparative Ethnographic Study of the Use of Mobile Information and Communication Technologies. *Atlantic Journal of Communication*, 14(4), pp. 191–210.

Koebler, J. (2016, 26 January). How a DIY Network Plans to Subvert Time Warner Cable's NYC Internet Monopoly. *Motherboard*. Available at: motherboard.vice.com/en_us/article/gv5qb4/how-a-diy-network-plans-to-subvert-time-warner-cables-nyc-internet-monopoly.

Konomi, S., Sasao, T., Hosio, S., and Sezaki, K. (2017). Exploring the Use of Ambient Wifi Signals to Find Vacant Houses. European Conference on Ambient Intelligence. In A. Braun, R. Wichert, and A. Māna, eds. *Ambient Intelligence. AmI 2017. Lecture Notes in Computer Science, vol. 10217*. Cham, Switzerland: Springer, pp. 130–5.

Kusiak, J. and Kacperski, W. (2012). Kiosks with Vodka and Democracy: *Civic Cafés* Between New Urban Movements and Old Social Divisions. In M. Grubbauer and J. Kusiak, eds. *Chasing Warsaw: Socio-Material Dynamics of Urban Change Since 1990*. Frankfurt: Campus Verlag, pp. 213–38.

La Habana. (2016, September). *La Habana Magazine*. Available at: www.lahabana.com/content/wp-content/uploads/2016/magazine/september/lahabana-magazine-sept2016.pdf.

Lacoma, T. (2020, 4 January). What Is Wi-Fi 6? Here's Everything You Need to Know. *TechCrunch*. Available at: www.digitaltrends.com/computing/what-is-wi-fi-6.

Lally, E. (2002). *At Home with Computers*. Oxford: Berg.

Lambert, A., McQuire, S., and Papastergiadis, N. (2013, August). *Free Wi-Fi and Public Space: The State of Australian Public Initiatives*. Melbourne: The University of Melbourne / Institute for a Broadband-Enabled Society (IBES). Available at: https://networkedsociety.unimelb.edu.au/__data/assets/pdf_file/0007/1661317/Free-Wi-Fi-and-Public-Space.pdf.

Lambert, A., McQuire, S., and Papastergiadis, N. (2014). Public Wi-Fi: Space, Sociality and the Social Good. *Australian Journal of Telecommunications and the Digital Economy*, 2(3), pp. 45.1–45.17.

Lampland, M. and Star, S. L. (2009). Reckoning with Standards. In: M. Lampland and S. L. Star, eds. *Standards and their stories: How*

Quantifying, Classifying, and Formalizing Practices Shape Everyday Life. New York: Cornell University Press, pp. 3–34.

Langley, N. (2003, 23 June). The Demise of the Warchalkers. *Computer Weekly*. Available at: www.computerweekly.com/feature/ The-demise-of-the-warchalkers.

Lap-Top Computers Gain Stature as Power Grows. (1987, 12 April). *Daily News of Los Angeles (CA)*. Available at: https://bit.ly/364Sj1S.

Larkin, B. (2013). The Politics and Poetics of Infrastructure. *Annual Review of Anthropology*, 42, pp. 327–43.

Laurier, E. (2008a). Drinking up Endings: Conversational Resources of the Café. *Language & Communication*, 28(2), pp. 165–81.

Laurier, E. (2008b). How Breakfast Happens in the Café. *Time & Society*, 17(1), pp. 119–34.

Laurier, E. and Philo, C. (2006a). Cold Shoulders and Napkins Handed: Gestures of Responsibility. *Transactions of the Institute of British Geographers*, 31(2), pp. 193–207.

Laurier, E. and Philo, C. (2006b). Possible Geographies: A Passing Encounter in a Café. *Area*, 38(4), pp. 353–63.

Laurier, E. and Philo, C. (2007). 'A Parcel of Muddling Muckworms': Revisiting Habermas and the English Coffee-Houses. *Social & Cultural Geography*, 8(2), pp. 259–81.

Laurier, E., Whyte, A., and Buckner, K. (2001). An Ethnography of a Neighbourhood Café: Informality, Table Arrangements and Background Noise. *Journal of Mundane Behaviour*, 2(2), pp. 195–232.

Lave, J. and Wenger, E. (1991). *Situated Learning: Legitimate Peripheral Participation*. Cambridge University Press.

Lehnardt, J. (2012, 26 July). Freifunk Statt Angst. *Writing*. Available at: https://writing.jan.io/2012/07/26/freifunk-statt-angst.html.

Lehr, W. and McKnight, L. W. (2003). Wireless Internet Access: 3G vs. WiFi? *Telecommunications Policy*, 27, pp. 351–70.

Lemstra, W. and Hayes, V. (2008). Unlicensed Innovation: The Case of Wi-Fi. *Competition and Regulation in Network Industries*, 9(2), pp. 135–71.

Lemstra, W., Hayes, V., and Groenewegen, J., eds. (2011). *The Innovation Journey of Wi-Fi: The Road to Global Success*. Cambridge University Press.

Lenczner, M. (2005, 31 October). Wireless Portals with Wifidog. *Linux Journal*. Available at: www.linuxjournal.com/article/8352.

Lennar. (2019). Lennar Launches World's First Wi-Fi CERTIFIED Home Design, Activation and Support by Amazon. Available at: www.lennar.com/wifi-certified.

Leong, E. (2013). Collecting Knowledge for the Family: Recipes, Gender and Practical Knowledge in the Early Modern English Household. *Centaurus: An International Journal of the History of Science and its Cultural Aspects*. Available at: https://onlinelibrary.wiley.com/doi/full/10.1111/1600-0498.12019.

Lewallen, C. M. and Seid, S., eds. (2004). *Ant Farm 1968–1978*. University of California Press / Berkeley Art Museum and Pacific Film Archive.

Lewis, T. (2020). *Digital Food: From Paddock to Platform*. London: Bloomsbury.

Lillywhite, B. (1963). *London Coffee Houses: A Reference Book of Coffee Houses of the Seventeenth, Eighteenth and Nineteenth Centuries*. London: G. Allen and Unwin.

LinkLabs. (2018, 1 February). WiFi's Future: Examining 802.11ad, 802.11ah HaLow (& Others). Available at: www.link-labs.com/blog/future-of-wifi-802-11ah-802-11ad.

Links, C. (2003). The Spirit of Wi-Fi: Where It Came From, Where It Is Today, and Where It Is Going. *Quorvo*. Available at: https://quorvo.com/resources/d/the-spirit-of-wifi-cees-links-2003.

Lobato, R. and Thomas, J. (2015). *The Informal Media Economy*. Cambridge: Polity.

Lofland, L. H. (1989a). The Morality of Urban Public Life: The Emergence and Continuation of a Debate. *Places*, 6(Fall), pp. 18–23.

Lofland, L. H. (1989b). *The Public Realm: Exploring the City's Quintessential Social Territory*. New Brunswick: Transaction Publishers.

Luque-Ayala, A. and Marvin, S. (2015). Developing a Critical Understanding of Smart Urbanism? *Urban Studies*, 52(12), pp. 2105–16.

Maccari, L. and Lo Cigno, R. (2015). A Week in the Life of Three Large Wireless Community Networks. *Ad Hoc Networks*, 24, pp. 175–90.

Mackenzie, A. (2005). Untangling the Unwired: Wi-Fi and the Cultural Inversion of Infrastructure. *Space & Culture*, 8(3), pp. 269–85.

Mackenzie, A. (2006a). From Café to Park Bench: Wi-Fi and Technological Overflows in the City. In M. Sheller and J. Urry, eds. *Mobile Technologies of the City*. London: Routledge, pp. 137–51.

Mackenzie, A. (2006b). Innumerable Transmissions: Wi-Fi from Spectacle to Movement. *Information, Communication & Society*, 9(6), pp. 781–802.

Mackenzie, A. (2007). Wireless Networks and the Problem of Over-Connectedness. *Media International Australia*, 125, pp. 94–105.

Mackenzie, A. (2010). *Wirelessness: Radical Empiricism in Network Cultures*. Cambridge, MA: MIT Press.

Madasu, V. (2016, 22 December). An Introduction to Smart Homes and IoT. *Medium*. Available at: https://medium.com/future-spaces/an-introduction-to-smart-homes-and-iot-f3cc477b52bc.

Manning, P. (2013). The Theory of the Café Central and the Practice of the Café Peripheral: Aspirational and Abject Infrastructures of Sociability on the European Periphery. In A. Tjora and G. Scambler, eds. *Café Society*. New York: Palgrave Macmillan, pp. 43–65.

Marcus, M. J. (2009). Wi-Fi and Bluetooth: The Path from Carter and Reagan-era Faith in Deregulation to Widespread Products Impacting Our World. *Info*, 11(5), pp. 19–35.

Marcus, M. J. (n.d.). Commercial Spread Spectrum Background. IEEE 802.11 Working Group. Available at: www.ieee802.org/802_tutorials/04-March/Revised-802-SS-talk.pdf.

MarketWatch. (2014, 18 June). iGR Study: Almost 98% of Broadband Data Use in U.S. Households to Be on WiFi Devices by 2018. Available at: www.marketwatch.com/press-release/new-igr-study-forecasts-that-almost-98-percent-of-broadband-data-use-in-us-households-will-be-on-wifi-devices-by-2018-2014-06-18.

Marr, B. (2020, 13 January). The 5 Biggest Smart Home Trends in 2020. *Forbes*. Available at: www.forbes.com/sites/bernardmarr/2020/01/13/the-5-biggest-smart-home-trends-in-2020/#271c15e8389b.

Martínez, A. G. (2017, 26 July). Inside Cuba's D.I.Y. Internet Revolution. *Wired*. Available at: www.wired.com/2017/07/inside-cubas-diy-internet-revolution.

Martinussen, E. S. (2011, 22 February). Immaterials: Light Painting WiFi. YOUrban. Available at: http://yourban.no/2011/02/22/immaterials-light-painting-wifi.

Mattern, S. (2013, November). Methodolatry and the Art of Measure: The New Wave of Urban Science. *Places*. Available at: https://placesjournal.org/article/methodolatry-and-the-art-of-measure/?cn-reloaded=1.

Mattern, S. (2017). *Code + Clay... Data + Dirt: Five Thousand Years of Urban Media*. Minneapolis: University of Minnesota Press.

Mattern, S. (2019, 8 July). Networked Dream Worlds. *Real Life*. Available at: https://reallifemag.com/networked-dream-worlds.

McConnell, C. and Straubhaar, J. (2015). Contextualizing Open Wi-Fi Network Use with Multiple Capitals. *Studies in Media and Communications*, 10, pp. 205–32.

McCormack, D. P. (2017). Elemental Infrastructures for Atmospheric

Media: On Stratospheric Variations, Value and the Commons. *Environment and Planning D: Society and Space*, 35(3), pp. 418–37.

McCosker, A. and Wilken, R. (2012). Café Space, Communication, Creativity, and Materialism. *M/C Journal*, 15(2). Available at: http://journal.media-culture.org.au/index.php/mcjournal/article/view/459.

McCosker, A., Vivienne, S., and Johns, A., eds. (2016). *Negotiating Digital Citizenship: Control, Contest and Culture*. London: Rowman & Littlefield International.

McQuire, S. (2008). *The Media City: Media, Architecture and Urban Space*. London: Sage.

McShane, I., Wilson, C., and Meredith, D. (2014). Broadband as Civic Infrastructure: The Australian Case. *Media International Australia*, 151, pp. 127–36.

Meese, J., Wilken, R., Nansen, B., and Arnold, M. (2015). Entering the Graveyard Shift: Disassembling the Australian TiVo. *Television & New Media*, 16(2), pp. 165–79.

Merriman, P. and Jones, R. (2017). Nations, Materialities and Affects. *Progress in Human Geography*, 41(5), pp. 600–17.

Metcalfe, B. (1993). From the Ether: Wireless Computing Will Flop – Permanently. *InfoWorld*, 16 August, p. 48.

Metz, E. (2017, 28 February). Some Cafes Are Banning Wi-Fi to Encourage Conversation. *BBC News*. Available at: www.bbc.com/worklife/article/20170224-some-cafes-are-banning-wi-fi-to-encourage-conversation.

Middleton, C. (2007, 15 August). A Framework for Investigating the Value of Public Wireless Networks. SSRN. Available at: https://ssrn.com/abstract=2118153.

Middleton, C. and Crow, B. (2008). Building Wi-Fi Networks for Communities: Three Canadian Cases. *Canadian Journal of Communication*, 33, pp. 419–41.

Mitchell, W. J. (2003). *Me++: The Cyborg Self and the Networked City*. Cambridge, MA: MIT Press.

Montgomery, John. (1997). Café Culture and the City: The Role of Pavement Cafés in Urban Public Social Life. *Journal of Urban Design*, 2(1), pp. 83–102.

Morley, D. (1986). *Family Television: Cultural Power and Domestic Leisure*. London: Comedia Publishing Group.

Morley, D. (2003). What's 'Home' Got to Do with It? Contradictory Dynamics in the Domestication of Technology and the Dislocation of Domesticity. *European Journal of Cultural Studies*, 6(4), pp. 435–58.

Morley, D. (2017). *Communications and Mobility: The Migrant, the Mobile Phone, and the Container Box*. Oxford: Wiley Blackwell.

Morrison, G. (2016, 23 February). Wireless HD Video is Here, So Why Do We Still Use HDMI Cables? CNet. Available at: www.cnet.com/news/wireless-hd-video-is-here-so-why-do-we-still-use-hdmi-cables.

Mossberger, K., Tolbert, C. J., and McNeal, R. S. (2007). *Digital Citizenship: The Internet, Society, and Participation*. Cambridge, MA: MIT Press.

Mullin, J. (2012a, 5 April). How the Aussie Government 'Invented WiFi' and Sued Its Way to $430 Million. Ars Technica. Available at: https://arstechnica.com/tech-policy/2012/04/how-the-aussie-government-invented-wifi-and-sued-its-way-to-430-million/2.

Mullin, J. (2012b, 6 April). Responses and Clarifications on the CSIRO Patent Lawsuits. Ars Technica. Available at: https://arstechnica.com/tech-policy/2012/04/op-ed.

Mumford, L. (1979 [1961]). *The City in History*. Harmondsworth, Middx: Penguin.

Nansen, B. (2019). Discourses, Dispositifs, and Dispositions in Young Children's Mobile Media Use. In L. Green, D. Holloway, K. Stevenson, and K. Jaunzems, eds. *Digitising Early Childhood*. Newcastle upon Tyne: Cambridge Scholars Publishing, pp. 28–45.

Nansen, B., Arnold, M., Gibbs, M., and Davis, H. (2011). Dwelling with Media Stuff: Latencies and Logics of Materiality in Four Australian Homes. *Environment and Planning D: Society and Space*, 29, pp. 693–715.

Navarro, L., Freitag, F., Baig, R., and Roca, R. (2016). A Commons-Oriented Framework for Community Networks. In L. Belli, ed. *Community Connectivity: Building the Internet from Scratch. Annual Report of the UN IGF Dynamic Coalition on Community Connectivity*. Rio de Janeiro, Brazil: FGV Direito Rio, pp. 55–92.

Navarro, L., Maccari, L., and Lo Cigno, R. (2018). At the Limits of the Network: Technology Options for Community. In: APC-IDRC, ed. *Global Information Society Watch 2018: Community Networks*. Johannesburg, South Africa, and Ottawa, Canada: Association for Progressive Communications (APC) / International Development Research Centre (IDRC), pp. 13–20.

Negroponte, N. (1995). *Being Digital*. Cambridge, MA: MIT Press.

NetSpot. (2019). What Is Wi-Fi RTT? NetSpot. Available at: www.netspotapp.com/what-is-wifi-rtt.html.

Niezen, R. (2014). Gabriel Tarde's Publics. *History of the Human Sciences*, 27(2), pp. 41–59.

Nordic Semiconductor. (2012, 26 January). A Short History of Spread Spectrum. *EE Times*. Available at: www.eetimes.com/document. asp?doc_id=1279374#.

Notaker, H. (2017). *A History of Cookbooks: From Kitchen to Page over Seven Centuries*. Berkeley: University of California Press.

NovaTech SE. (2016, 12 November). Google Wifi Commercial. YouTube. Available at: https://youtu.be/PXh8lpHm16k.

NYCwireless. (n.d.) About. Available at: https://nycwireless.wordpress. com/about-us.

O'Brien, C. (2009, 30 September). Is Free Wi-Fi a Good Deal for Coffee Shops? Phys.Org. Available at: https://phys.org/news/2009-09-free-wi-fi-good-coffee.html.

O'Neil, D. (2002). Assessing Community Informatics: A Review of Methodological Approaches for Evaluating Community Networks and Community Technology Centers. *Internet Research*, 12(1), pp. 76–102.

Oldenburg, R. (1989). *The Great Good Place: Cafes, Coffee Shops, Community Centers, Beauty Parlors, General Stores, Bars, Hangouts, and How They Get You Through the Day*. New York: Paragon House.

Oldenburg, R. (2013). The Café as a Third Place. In A. Tjora and G. Scambler, eds. *Café Society*. New York: Palgrave Macmillan, pp. 7–21.

Palmer, D. (2010). Emotional Archives: Online Photo Sharing and the Cultivation of the Self. *Photographies*, 3(2), pp. 155–71.

Parekh, J. (2017, 11 May). WiFi's Evolving Role in IoT. *Networkworld*. Available at: www.networkworld.com/article/3196191/wifi-s-evolving-role-in-iot.html.

Park, N. and Lee, K. M. (2010). Wireless Cities: Local Governments' Involvement in the Shaping of Wi-Fi Networks. *Journal of Broadcasting and Electronic Media*, 54(3), pp. 425–42.

Parks Associates. (2018). 76% of North American Broadband Households Use Wi-Fi as Their Primary Connection Technology. Available at: www.parksassociates.com/blog/article/pr-01242018.

PC Powerplay. (2001, May). 3Com Wireless LAN Starter Pack. Available at: https://archive.org/details/PCPowerplay-060-2001-05/page/n99/mode/2up/search/Wi-Fi?q=Wi-Fi.

Perng, S.-Y. (2015). Performing Tasks with Wi-Fi Signals in Taipei. *Space & Culture*, 18(3), pp. 285–97.

Pertierra, A. C. (2012). If They Show *Prison Break* in the United States on a Wednesday, by Thursday It Is Here: Mobile Media Networks in Twenty-First Century Cuba. *Television and New Media*, 13(5), pp. 399–414.

Plantin, J.-C., Lagoze, C., Edwards, P. N., and Sandvig, C. (2018). Infrastructure Studies Meet Platform Studies in the Age of Google and Facebook. *New Media & Society*, 20(1), pp. 293–310.

Poblet, M. (2018). Distributed, Privacy-Enhancing Technologies in the 2017 Catalan Referendum on Independence: New Tactics and Models of Participatory Democracy. *First Monday*, 23(12). Available at: https://doi.org/10.5210/fm.v23i12.9402.

Potts, J. (2014). Economics of Public WiFi. *Australian Journal of Telecommunications and the Digital Economy*, 2(1), pp. 20.1–20.9.

Powell, A. (2008). WiFi Publics. *Information, Communication & Society*, 11(8), pp. 1068–88.

Powell, A. (2009, 21 January). Wi-Fi as Public Utility or Public Park? Metaphors for Planning Local Communication Infrastructure. SSRN. Available at: http://dx.doi.org/10.2139/ssrn.1330913.

Powell, A. (2012). Wi-Fi Publics: Defining Community and Technology at Montréal's Île Sans Fil. In A. Clement, M. Gurstein, and G. Longford, eds. *Connecting Canadians: Investigations in Community Informatics*. Edmonton: Athabasca University Press, pp. 202–17.

Powell, A. and Shade, L. R. (2006). Going Wi-Fi in Canada: Municipal and Community Initiatives. *Government Information Quarterly*, 23(3–4), pp. 381–403.

Powell, A. and Shade, L. R. (2012). Community and Municipal Wi-Fi Initiatives in Canada. In A. Clement, M. Gurstein, and G. Longford, eds. *Connecting Canadians: Investigations in Community Informatics*. Edmonton: Athabasca University Press, pp. 183–201.

Press, L. (2011). The Past, Present and Future of the Internet in Cuba. In *Cuba in Transition, Volume 21. Papers and Proceedings of the Twenty-First Meeting of the Association for the Study of the Cuban Economy (ASCE), Miami, August 4–6*. Available at: www.ascecuba.org/c/wp-content/uploads/2014/09/v21-press.pdf.

Puregger, A. (2018, 30 July). Are WiFi Networks Ready for Smart Cities? TechRadar. Available at: www.techradar.com/au/news/are-wifi-networks-ready-for-smart-cities.

Rabie, K. (2019, 26 June). 5G Orchestration – The Missing Brick [blog]. Netmanias. Available at www.netmanias.com/en/post/blog/14325/5g/5g-orchestration-the-missing-brick.

Radio Shack. (2001). Connecting People Places and Possibilities. Available at: www.radioshackcatalogs.com/catalogs/2001-a.

Radio Shack. (2003–4). Everyday Needs: Parts, Batteries & Accessories. 2003–2004 Reference Guide. Available at: www.radioshackcatalogs.com/catalogs/2003-04.

Radio Shack. (2010). 2010 Assortment Book. Available at: www.radioshackcatalogs.com/html/2010.

Reddigari, M. (2019, 13 March). How to Set Up Free WiFi for Customers. Microsoft. Available at: www.microsoft.com/en-us/microsoft-365/growth-center/resources/setup-free-wifi-customers.

Rennie, E., Hogan, E., Gregory, R., Crouch, A., Wright, A., and Thomas, J. (2016). *Internet on the Outstation: The Digital Divide and Remote Aboriginal Communities*. Amsterdam: Institute for Network Cultures.

Rennie, E., Yunkaporta, T., and Holcombe-James, I. (2019). Killswitch. Disconnect. Melbourne: RMIT University. [podcast] Available at: www.indigitube.com.au/album/5d3f91d6442def54b8422c8f.

Reply All. (2016, 4 February). #53 in the Desert. Gimlet Media. [podcast]. Available at: https://gimletmedia.com/shows/reply-all/n8hodm.

Richards, D. (2018, 22 October). 5G vs Wi Fi 6, Why Retailers Will Be the Big Winners. Channelnews. Available at: www.channelnews.com.au/5g-vs-wi-fi-6-why-retailers-will-be-the-big-winners.

Richtel, M. (2004, 2 December). Technology: Pennsylvania Limits Cities in Offering Net Access. *New York Times*. Available at: www.nytimes.com/2004/12/02/business/technology/technology-pennsylvania-limits-cities-in-offering-net.html.

Rittner, L., Haine, W. S., and Jackson, J. H. (2016). *The Thinking Space: The Café as a Cultural Institution in Paris, Italy and Vienna*. London: Routledge.

Roberts, H., and Wong, G. (2019). The Evolution of Wi-Fi. Available at: www.calix.com/content/dam/calix/marketing-documents/public/connexions_2019/The-Evolution-of-WiFi.pdf.

Robinson, L., Schulz, J., Khilnani, A., et al. (2020). Digital Inequalities in Time of Pandemic: COVID-19 Exposure Risk Profiles and New Forms of Vulnerability. *First Monday*, 25(7). Available at: https://firstmonday.org/ojs/index.php/fm/article/download/10845/9563, DOI: http://dx.doi.org/10.5210/fm.v25i7.10845.

Robinson, M. (2015, 14 November). This Charity Wants to Turn Homeless People into WiFi Hotspots. *Business Insider*. Available at: www.businessinsider.com.au/wifi-4-life-turns-homeless-people-into-internet-hotspots-2015-11?r=US&IR+T.

Rose, N. (1999). *Politics of Freedom: Reframing Political Thought*. Cambridge University Press.

Rosenberg, D. (2020, 10 January). 5G Is About to Change the World in Ways We Can't Even Imagine Yet. World Economic Forum. Available at: www.weforum.org/agenda/2020/01/5g-is-about-to-change-the-world-in-ways-we-cant-even-imagine-yet.

Sadowski, J. (2019). When Data Is Capital: Datafication, Accumulation, and Extraction. *Big Data & Society*, January–June, pp. 1–12. Available at: https://doi.org/10.1177/2053951718820549.

Sadowski, J. and Bendor, R. (2019). Selling Smartness: Corporate Narratives and the Smart City as a Sociotechnical Imaginary. *Science, Technology, & Human Values*, 44(3), pp. 540–63.

Sadowski, J. and Pasquale, F. (2015). The Spectrum of Control: A Social Theory of the Smart City. *First Monday*, 20(7). Available at: https://doi.org/10.5210/fm.v20i7.5903.

San Francisco Call. (1912). To Check Wireless Anarchy. *San Francisco Call*, 7 July, p. 22.

Sandbye, M. (2014). Looking at the Family Photo Album: A Resumed Theoretical Discussion of How and Why. *Journal of Aesthetics and Culture*, 6(1). Available at: https://doi.org/10.3402/jac.v6.25419.

Sandvig, C. (2004). An Initial Assessment of Cooperative Action in Wi-Fi Networking. *Telecommunications Policy*, 28, pp. 579–602.

Sandvig, C. (2012). What Are Community Networks an Example of? A Response. In A. Clement, M. Gurstein, and G. Longford, eds. *Connecting Canadians: Investigations in Community Informatics*. Edmonton: Athabasca University Press, pp. 133–40.

Sandvik, K., Thorhauge, A. M., and Valtysson, B., eds. (2016). *The Media and the Mundane: Communication Across Media in Everyday Life*. Göteborg, Sweden: NORDICOM.

Santo, B. (2017, 5 December). Wi-Fi versus 5G? Nope, It's Both. EDN Network. Available at: www.edn.com/electronics-blogs/5g-waves/4459120/Wi-Fi-versus-5G--Nope--it-s-both.

Sawhney, H. (2005). Wi-Fi Networks and the Reorganization of Wireline-Wireless Relationship. In R. Ling and P. E. Pedersen, eds. *Mobile Communication: Re-negotiation of the Social Sphere*. London: Springer-Verlag, pp. 45–61.

Scales, W. C. (1980, December). *Potential Use of Spread Spectrum Techniques in Non-Government Applications*. McLean, VI: The Mitre Corporation.

Scholtz, R. A. (1982, May). The Origins of Spread-Spectrum Communications. *IEEE Communications Magazine*, 5, pp. 822–54.

Schwartz, M. and Abramson, N. (2009, December). The AlohaNet – Surfing the Wireless Data. *IEEE Communications Magazine*, 47(12), pp. 21–5.

Schwarze, C. L. (2018). We Want Wi-Fi: The FCC's Intervention in Municipal Broadband Networks. *Washington University Journal of Law & Policy*, 56, pp. 199–220.

Searls, D. (2003, 1 September). Linux Makes Wi-Fi Happen in New York City. *Linux Journal*. . www.linuxjournal.com/article/6897.

Sennett, R. (1974). *The Fall of Public Man*. Cambridge University Press.

Shaffer, G. (2007). Frame-Up: An Analysis of Arguments For and Against Municipal Wireless Initiatives. *Public Works Management & Policy*, 11(3), pp. 204–16.

Shankland, S. (2018, 3 October). Here Come Wi-Fi 4, 5, and 6 in Plan to Simplify 802.11 Networking Names. CNet. Available at: www.cnet.com/news/wi-fi-alliance-simplifying-802-11-wireless-network-tech-names.

Shepherd, C., Arnold, M., Bellamy, C., and Gibbs, M. (2007). The Material Ecologies of Domestic ICTs. *Electronic Journal of Communication / La Revue Électronique de Communication*, 17(1–2). Available at: www.cios.org/EJCPUBLIC/017/1/01712.HTML.

Silverstone, R. (1994). *Television and Everyday Life*. London: Routledge.

Silverstone, R. and Haddon, L. (1996). Design and the Domestication of Information and Communication Technologies: Technical Change and Everyday Life. In R. Mansell and R. Silverstone, eds. *Communication by Design: The Politics of Information and Communication Technologies*. Oxford University Press, pp. 44–74.

Silverstone, R. and Hirsch, E. (1992). Introduction. In R. Silverstone and E. Hirsch, eds. *Consuming Technologies: Media and Information in Domestic Spaces*. London: Routledge, pp. 1–11.

Silverstone, R., Hirsch, E., and Morley, D. (1994). Information and Communication Technologies and the Moral Economy of the Household. In: R. Silverstone and E. Hirsch, eds. *Consuming Technologies: Media and Information in Domestic Spaces*. London: Routledge, pp. 15–31.

Singel, R. (2011, 15 December). U.S.-funded Internet Liberation Project Finds Perfect Test Site: Occupy D.C. *Wired*. Available at: www.wired.com/2011/12/internet-suitcase-dc.

Singer, M. (2005, 3 June). PC Milestone – Notebooks Outsell Desktops. CNet. Available at: www.cnet.com/news/pc-milestone-notebooks-outsell-desktops.

Smithers, R. (2007, 7 October). McDonald's to Offer Free Wi-Fi in Restaurants. *The Guardian*. Available at: www.theguardian.com/technology/2007/oct/06/internet.

Söderström, O., Paasche, T., and Klauser, F. (2014). Smart Cities as Corporate Storytelling. *City*, 18(3), pp. 307–20.

Song, S., Rey-Moreno, C., Esterhuysen, A., Jensen, M., and Navarro, L. (2018). Introduction: The Rise and Fall and Rise of Community

Networks. In: APC-IDRC, ed. *Global Information Society Watch 2018: Community Networks*. Johannesburg, South Africa, and Ottawa, Canada: Association for Progressive Communications (APC) / International Development Research Centre (IDRC), pp. 7–10.

Spigel, L. (1992). *Make Room for TV: Television and the Family Ideal in Postwar America*. University of Chicago Press.

Spigel, L. (2005). Designing the Smart House: Posthuman Domesticity and Conspicuous Production. *European Journal of Cultural Studies*, 8(4), pp. 403–26.

Spigel, L. (2010). Smart Homes: Digital Lifestyles Practiced and Imagined. In J. Gripsrud, ed. *Relocating Television: Television in the Digital Context*. London: Routledge, pp. 238–56.

Star, S. L. (2010). This Is Not a Boundary Object: Reflections on the Origins of a Concept. *Science, Technology & Human Values*, 35(5), pp. 601–17.

Star, S. L. and Griesemer, J. R. (1989). Institutional Ecology, 'Translations' and Boundary Objects: Amateurs and Professionals in Berkeley's Museum of Vertebrate Zoology, 1907–39. *Social Studies of Science*, 19, pp. 387–420.

Star, S. L. and Ruhleder, K. (1996). Steps Towards an Ecology of Infrastructure: Design and Access for Large Information Spaces. *Information Systems Research*, 7(1), pp. 111–34.

Stenseth, B. (2013). Heart of Urbanism. The Café: a Chapter of Cultural History. In A. Tjora and G. Scambler, eds. *Café Society*. New York: Palgrave Macmillan, pp. 23–42.

Strain, J. D. (2003). Households as Morally Ordered Communities: Explorations in the Dynamics of Domestic Life. In R. Harper, ed. *Inside the Smart Home*. London: Springer-Verlag, pp. 41–62.

Streeter, T. (1994). Selling the Air: Property and the Politics of US Commercial Broadcasting. *Media, Culture & Society*, 16, pp. 91–116.

Strengers, Y. and Kennedy, J. (2020). *The Smart Wife: Why Siri, Alexa, and Other Smart Home Devices Need a Feminist Reboot*. Cambridge, MA: MIT Press.

Tate, R. (2012, 2 November). Why Facebook Might Get into the Free Wi-Fi Racket. *Wired*. Available at: www.wired.com/business/2012/11/facebook-wifi.

Taylor, G., Middleton, C., and Goodrick, P. (2015). Getting Communications Policy out of the Telephone Booth. *Options Politiques*, May–June, pp. 46–8.

Telecom Advisory Services. (2018, October). The Economic Value of Wi-Fi: A Global View (2018 and 2023). Wi-Fi Alliance. Available at: www.wi-fi.org.

Telecomlead. (2016, 13 July). TRAI Gears Up for Public Wi-Fi Networks in India. Available at: www.telecomlead.com/broadband/trai-gears-public-wi-fi-networks-india-69908.

Telstra. (2020). Telstra 5G Wi-Fi Pro. Available at: www.telstra.com.au/internet/mobile-broadband/telstra-5G-wi-fi-pro.

Thomas, J., Barraket, J., Wilson, C. K., Rennie, E., Ewing, S., and MacDonald, T. (2019). *Measuring Australia's Digital Divide: The Australian Digital Inclusion Index 2019*. Melbourne: RMIT University and Swinburne University of Technology, for Telstra.

Tjora, A. (2013). Communal Awareness in the Urban Café. In A. Tjora and G. Scambler, eds. *Café Society*. New York: Palgrave Macmillan, pp. 103–26.

Torrens, P. M. (2008). Wi-Fi Geographies. *Annals of the Association of American Geographers*, 98(1), pp. 59–84.

Townsend, A. (2014, 18 November). After 4,984 Days, NYC Finally Has Free WiFi, Everywhere. Soon. Medium. Available at: https://medium.com/@anthonymobile/after-4-948-days-nyc-finally-has-free-wifi-everywhere-soon-d4376247373.

TRAI. (2016, 13 July). *Consultation Paper on Proliferation of Broadband through Public Wi-Fi Networks*. Telecom Regulatory Authority of India. Available at: https://main.trai.gov.in/consultation-paper-proliferation-broadband-through-public-wi-fi-networks.

United Nations International Children's Emergency Fund. (2020). *Covid-19: Are Children Able to Continue Learning during School Closures? A Global Analysis of the Potential Reach of Remote Learning Policies using Data from 100 Countries*. New York: UNICEF.

Urban Omnibus. (2013, 25 September). Local Connections: The Red Hook WiFi Project. Available at: https://urbanomnibus.net/2013/09/red-hook-wifi.

Urry, J. (2016). *What Is the Future?* Cambridge: Polity.

Vaarzon-Morel, P. (2014). Pointing the Phone: Transforming Technologies and Social Relations among Warlpiri. *The Australian Journal of Anthropology*, 25(2), pp. 239–55.

Van den Broeck, W., Pierson, J., and Lievens, B. (2008). Confronting video-on-demand with television viewing practices. In J. Pierson, E. Mante-Meijer, E. Loos, and B. Sapio, eds. *Innovating for and by Users*. Brussels: COST, pp. 13–26.

Van Grove, J. (2013, 2 October). Why Facebook Is Giving Out Free Wi-Fi for Check-ins. CNet. Available at: www.cnet.com/news/why-facebook-is-giving-out-free-wi-fi-for-check-ins.

Van Oost, E., Verhaegh, S., and Oudshoorn, N. (2009). From

Innovation Community to Community Innovation: User-initiated Innovation in Wireless Leiden. *Science, Technology & Human Values*, 34(2), pp. 182–205.

Vanolo, A. (2014). Smartmentality: The Smart City as Disciplinary Strategy. *Urban Studies*, 51(5), pp. 883–98.

Venegas, C. (2010). *Digital Dilemmas: The State, the Individual, and Digital Media in Cuba*. New Brunswick, NJ: Rutgers University Press.

von Hippel, E. (2017). *Free innovation*. Cambridge, MA: MIT Press.

Vos, E. (2009, 28 September). St Cloud Shuts Down Free Citywide WiFi Service. MuniWireless. Available at: http://muniwireless.com/2009/09/28/st-cloud-shuts-down-free-citywide-wifi-service.

Waddell, K. (2018, 4 May). The Unbelievably Teched-Out Houses of Smart-Home Obsessives. *New York Times Magazine*. Available at: https://nymag.com/intelligencer/smarthome/extreme-makeover-smart-home-edition.html.

Wagstaff, J. (2004, 4 November). Home Is Where the Hand-held Is. *Far Eastern Economic Review*, pp. 38–41.

Wall, M. (2020, 28 January). What Is 5G and What Will It Mean for You? *BBC News*. Available at: www.bbc.com/news/business-44871448.

Walters, P. and Broom, A. (2013). The City, the Café, and the Public Realm in Australia. In A. Tjora and G. Scambler, eds. *Café Society*. New York: Palgrave Macmillan, pp. 185–205.

Wang, M., Liao, F. H., Lin, J., Huang, L. Gu, C. and Wei, Y. D. (2016). The Making of a Sustainable Wireless City? Mapping Public Wi-Fi Access in Shanghai. *Sustainability*, 8(111). Available at: https://doi.org/10.3390/su8020111.

Wark, M. (1994). Third Nature. *Cultural Studies*, 8(1), pp. 115–32.

WBA. (2017, 13 December). *WBA Annual Industry Report 2017/18*. Wireless Broadband Alliance. Available at: www.wballiance.com/wp-content/uploads/2017/12/WBA-Industry-Report-2017.pdf.

WBA. (2018, October). *WBA Annual Industry Report 2019*. Wireless Broadband Alliance. Available at: www.wballiance.com.

Wearden, G. (2002, 6 September). Warchalking a Map for Drive-by Spammers. ZDNet. Available at: www.zdnet.com/article/warchalking-a-map-for-drive-by-spammers.

Webb, W. (2011). The Networked Home: The Way of the Future or a Vision Too Far? In R. Harper, ed. *Connected Homes: The Future of Domestic Life*. London: Springer, pp. 19–28.

Weinreich, A. (2017, 18 December). The Future of the Smart Home: Smart Homes & IoT: A Century in the Making. *Forbes*.

Available at: www.forbes.com/sites/andrewweinreich/2017/12/18/the-future-of-the-smart-home-smart-homes-iot-a-century-in-the-making/#ca97e2f57ac2.

Werbach, K. (2004). Supercommons: Toward a Unified Theory of Wireless Communication. *Texas Law Review*, 82, pp. 863–973.

White, M. (2012). *Coffee Life in Japan*. Los Angeles: University of California Press.

Why I Founded Melbourne Wireless. (n.d.). Steve's Blog [blog]. Available at: www.crc.id.au/why-i-founded-melbourne-wireless.

Wi-Fi Alliance. (2017, January). Wi-Fi Alliance Publishes 7 for '17 Wi-Fi Predictions. Available at: www.wi-fi.org/news-events/newsroom/wi-fi-alliance-publishes-7-for-17-wi-fi-predictions.

Wi-Fi Alliance. (2019a). Wi-Fi Home Design. Available at: www.wi-fi.org/discover-wi-fi/wi-fi-home-design.

Wi-Fi Alliance. (2019b). WiFi® in 2019. Available at: www.wi-fi.org/news-events/newsroom/wi-fi-in-2019.

Wi-Fi Alliance. (2020a). Wi-Fi Alliance Brand Style Guide – June 2020. Available at: www.wi-fi.org/download.php?file=/sites/default/files/private/Wi-Fi_Alliance_Brand_Style_Guide_202006.pdf.

Wi-Fi Alliance. (2020b). Wi-Fi Is ... Available at: www.wi-fi.org/discover-wi-fi.

Wi-Fi Attendance. (2018, 27 August). How Will WiFi Networks Influence Smart City Development? Available at: www.wifiattendance.com/blog/wifi-networks-influence-smart-city-development.

Wigley, M. (1995). *White Walls, Designer Dresses: The Fashioning of Modern Architecture*. Cambridge, MA: MIT Press.

Wikipedia. (2017). Wireless Gateway. Available at: https://en.wikipedia.org/wiki/Wireless_gateway.

Wikipedia. (2019a). IEEE 802.11. Available at: https://en.wikipedia.org/wiki/IEEE_802.11.

Wikipedia. (2019b). Île Sans Fil. Available at: https://en.wikipedia.org/wiki/Île_Sans_Fil.

Wikipedia. (2019c). Mobile Data Offloading. Available at: https://en.wikipedia.org/wiki/Mobile_data_offloading.

Wikipedia. (2019d). Residential Gateway. Available at: https://en.wikipedia.org/wiki/Wireless_gateway.

Wikipedia. (2020a). ALOHAnet. Available at: https://en.wikipedia.org/wiki/ALOHAnet.

Wikipedia. (2020b). Google WiFi. Available at: https://en.wikipedia.org/wiki/Google_WiFi.

Wikipedia. (2020c). ISM Band. Available at: https://en.wikipedia.org/wiki/ISM_band.

Wilken, R. and Goggin, G. (2015). Locative Media – Definitions, Histories, Theories. In R. Wilken and G. Goggin, eds. *Locative Media*. New York: Routledge, pp. 1–19.

Wilken, R. (2019). Communication Infrastructures and the Contest over Location Positioning. *Mobile Media & Communication*, 7(3), pp. 341–61.

Wilken, R., Arnold, M., and Nansen, B. (2011). Broadband in the Home Pilot Study: Suburban Hobart. *Telecommunications Journal of Australia*, 61(1), pp. 5.1–5.16.

Wilken, R., Nansen, B., Arnold, M., Kennedy, J., and Gibbs, M. (2014). National, Local and Household Media Ecologies: The Case of Australia's National Broadband Network. *Communication, Politics & Culture*, 46(2), pp. 136–54. Available at: https://apo.org.au/node/38419.

Willett, R. (2017). Domesticating Online Games for Preteens – Discursive Fields, Everyday Gaming, and Family Life. *Children's Geographies*, 15(2), pp. 146–59.

Wilson, R. (2015, 20 October). Understanding the IoT Cloud and How It Will Change Things. *Electronics Weekly* Available at: www.electronicsweekly.com/news/understanding-the-iot-cloud-and-how-it-will-change-things-2015-10.

Wiman. (2020). Where May I Hook Up to Free Internet in Indonesia? Available at: www.wiman.me/indonesia.

Wired. (2004, 3 August). Wi-Fi Shootout in the Desert. Available at: www.wired.com/2004/08/wi-fi-shootout-in-the-desert.

Wyatt, T. (2018, 6 December). Cuba to Finally Give Citizens Internet Access on their Phones as Government Launches 3G Service. *Independent*. Available at: www.independent.co.uk/news/world/americas/cuba-internet-access-mobile-cell-phones-wifi-3g-connection-telecoms-a8667771.html.

Young, I. (1986). The Ideal of Community and the Politics of Difference. *Social Theory and Practice*, 12(1), pp. 1–26.

Zinn, J. O. and McDonald, D. (2018). Risk in the *New York Times* (1987–2014). In J. O. Zinn and D. McDonald, eds. *Risk in the New York Times (1987–2014): A Corpus-Based Exploration of Sociological Theories*. London: Palgrave Macmillan, pp. 81–136. Available at: https://doi.org/10.1007/978-3-319-64158-4.

Index

Page numbers in **bold** denote illustrations

3G mobile networks 6
4G 14
5G 8, 14, 22, 44, 76–7, 131, 147, 149
   and the smart city 136–8
802.11 standards 12–13, 14, 21, 22, 38
   802.11ac (Wi-Fi 5) 12, 14, 44, 67
   802.11ad (AD) 74–5
   802.11af (AF) 74
   802.11ah (HaLow) 74, 136
   802.11ax *see* Wi-Fi 6
   802.11b 12, 43
   802.11n (Wi-Fi 4) 44
   development of 39–40, 41

Aboriginals, Australian 46–7
access, and city Wi-Fi 131–5
AD standard (802.11ad) 74–5
advertisements 57–8
   Google Wifi 60
   Wi-Fi routers 60
AF (802.11af) standard 74
Air-Stream 90, 129
AirPort 12
Alibaba 149
Alliance for Affordable Internet 10
ALOHANet 18, 36, **37**
Amazon 149
Anderson, Chris 19
Andrejevic, Mark 126
Andrew Project 42
Android operating system 118

Ant Farm, House of the Century project 72
Antheil, George 35
Apple 4, 12, 16, 42, 117, 149
Apple AirPort base station 42, **43**
Apple iBook 4–5, 12, 42, 117, 145
Apple iPad 118
Apple iPhone 118, 144
appropriation (domestication approach) 56, 61–3
Arnall, Timo 108
Ars Technica 20–2
Arthur, Brian 16
AT&T 92, 148
Athens Metropolitan Wide Network 89–90
Atomic magazine 58
Austin (Texas) 101, 133
Australia 103, 127
   digital exclusion and Indigenous communities 46–8
   forest fires (2020) 1–2, 8, 9, 151
   Indigenous governance systems and Wi-Fi 103–6
   NG Media 98
   setting up of mobile connection units after fires 2, **3**
   Wi-Fi and the homeless 134
Australian Private Networks 48
Australian Communications and Media Authority 64

Index

Barns, Sarah 138
Barry, Charles 71
basic service set (BSS) colours 75, 76
Behrendt, Larissa 105–6
Bell Atlantic Mobile 38
Benkler, Yochai 92–3
Birdman, Jodie 70
Bluetooth 16, 52, 152
  tracing apps 152–3
Bly, Sara 65
Boggs, David 36
boosters 65
boundary objects 39
brand, Wi-Fi as a 12–16
Brazil 127
Briar (messaging app) 100
Broadcom 144
Broadreach 128
Busch, Lawrence 39

cafés 110, 119–26, 133, 139
  advantages of Wi-Fi for owners 123
  as enablers of communication and creativity 122
  and intimate anonymity 123
  role in formation of modern sociability 122
  roots and history 120–2
  significance of in urban life 120
  trade-offs to be made by owners 123–4
  ways of setting up / maintaining of Wi-Fi by owners 124–5
Canada 127
  Fredericton eZone 127, 128
  Île Sans Fil project (Montréal) 88–9, 101, 127, 128
  Canadian Research Alliance for Community Innovation and Networking (CRACIN) 97
Carnegie, Andrew 42

Carnegie Mellon University 18, 42
carrier sense multiple access (CSMA) 36
carrier sense multiple access with collision avoidance (CSMA-CA) 36
cash registers 38
CE Tips magazine 59–60
cellular broadband 8, 149
Cellular Digital Packet Data 38
cellular networks 2, 9, 42, 118, 137, 141, 146
cellular wireless services 6
central business districts (CBDs) 112
Centre for Appropriate Technology (CAT) 47
children, controlling ICT use of by parents 67–8
China, and 5G 147
China Mobile Communications Corporation (CMCC) 133
chips, Wi-Fi 144–5
Cincinnati Wi-Fi 133
Cisco
  Enterprise Networking Group 125
  Wi-Fi Opportunity Pyramid 129
cities
  and communication technologies 110, 111–16
  expansion of 112
  information infrastructure 114
  and public realm 120
  smart *see* smart cities
city Wi-Fi 108–39
  access and equity 131–5
  and cafés 110, 119–26, 133, 139
  concerns 110
  connectivity issues 113–14, 115, 117, 136
  as a contested space of possibility 111, 126–31, 135, 139

developments in laptops / mobile devices, and diffusion of 116–19, 139
and digital inclusion/exclusion 131–5
early problems encountered 129–30
factors fuelling acceleration of 110
growth of hotspots 119
homelessness and free 134
meaning 115
ownership models of wireless networks 128–9
patchwork quilt metaphor 114, 115
platformizing of 111, 116, 126, 138
politics and economics of 126–31, 139
smart cities 135–8
Coase, Ronald 32
coffee-houses (London) 120–1
commons infrastructures 101
Commonwealth Scientific and Industrial Research Organisation *see* CSIRO
Commotion Wireless system 99
communication networks, importance of 2
communication technologies and the city 110, 111–16
communities of practice 39–40
community, definition 83
community associations 93
community informatics 97–8
community infrastructure cooperatives 84–7
community innovation 95–6
community organizations, definition 83–4
community-owned wireless networks 128–9
community Wi-Fi networks 24, 82–107
characteristics 106–7
embedding of in various infrastructures / contractual arrangements 100–2
important functions of 107
Indigenous governance systems 103–6
and ISF (Île Sans Fil) 88–9, 101
limits of 100–3
mesh networks 98–100
number of 89–90
as open infrastructure 92–4
as process of innovation 94–7
rise of 87–90
social coordination resulting from 82, 84
and spectrum regulation 102–3
values 93
warchalking 90–2, 91
computer magazines
adverts for routers 60
articles on Wi-Fi 57–9
computing infrastructures 29
connected home 71–8
Connected Mobile Experiences (CMX) 125
connectivity 46
and city Wi-Fi 113–14, 115, 117, 136
and household Wi-Fi 52, 59, 60, 64, 65, 73, 74
improving of by network performance-enhancing products 65
consumer electronics magazines, articles on Wi-Fi 57, 59–60
contact tracing apps 152–3
conversion (domestication approach) 56–7, 69–70, 79
cookbooks 80
cordless microphones 42
Covid-19 pandemic 2–4, 8–9, 140, 151

Index

contact tracing apps 151–3
CSIRO (Commonwealth Scientific and Industrial Research Organisation) 18, 20–2
  patent dispute 20–2
Cuba
  and internet 85–6
  and SNET 86

de Camargo Penteado, Claudio Lui 127
De Filippi, Primavera 93
de Waal, Martijn 134–5
DEF CON hacking shoutout 96
Dell 42
Dethlefs, Jan 154
digital divide/inequality 7, 8–9, 131–2
  impact of Covid-19 pandemic on 10–11
  reducing of by free public Wi-Fi 131–2
digital exclusion 26, 45, 46–50, 98
  and Indigenous communities in Australia 46–8
  Sarawak (Malaysia) 48–9
digital inclusion 26, 46
  and free public Wi-Fi 132–4
direct sequence 35, 41, 43
Dobson, Wayne 25
domestic technology, Wi-Fi as key 51
domestic Wi-Fi *see* household Wi-Fi
domestication framework 23–4, 52, 53–71, 77, 78–80
  appropriation 56, 61–3
  conversion 56–7, 69–70, 79
  defining 55
  imagination 56, 57–61
  incorporation 56–7, 67–9
  objectification 56, 63–7
  regarding of home as privileged site 55–6

shallow 78
drones 142
Dunne, Anthony 113

Easyweb Digital 106
Economist, The 19
Edwards, Paul N. 45
equity, and city Wi-Fi 131–5
Ericsson Mobility Report 53
Ethernet 40
  first versions of 36–7
Ethos Global 104–5
extenders 65
Extreme Networks 136

Facebook 106, 125–6, 149
family life, Wi-Fi as integral to contemporary 51
FCC (Federal Communications Commission) 19, 31–2, 33, 35, 45, 51, 146, 148
  decision to release new block of spectrum for unlicensed use (2020) 146–7
  MITRE report (1980) 35
Ferris, Charles 32
finder apps 25
Firechat 100
First Nations Media Australia (FNMA) 104
Fischer, Claude S. 85
Fon 92, 135
Forlano, Laura 28–9, 130
Fredericton eZone (Canada) 127, 128
free innovation paradigm 95
free software communities 92, 93–4, 95
Freifunk Freedom Fighter Box 103
Freifunk Germany 90, 103
French Republic of Letters 120, 121
frequency division 30
frequency hopping 16, 35, 41, 43

Fuentes-Bautista, Martha 101
Fuller, Matthew 67

gaming 58
gaming consoles 117
Garcia, Laura 136
Garton, Andrew 102
Gates, Bill 72, 78
General Packet Radio Service (GPRS) 38
generational Wi-Fi 14
Germany
  Freifunk 90, 103
  Störerhaftung law (2017) 103
Giovanella, Federica 103
GitHub 94, 100
GNU General Public licence 88
Goggin, Gerard 52, 117, 118, 127, 129
Google 67, 116, 149
  Android operating system 118
Google Search 144
Google Wifi 50, 60, 61, 128
government(s) 19, 22, 30, 31, 82, 83, 101
GPS 16, 25
gradualism, Wi-Fi 144
Graham, Stephen and Marvin, Simon, *Telecommunications and the City* 112
Groenewegen, John 20, 39, 41, 45
Grubesic, Tony 133
guifi.net 90, 103
Guifi.net Foundation 101
Gurstein, Michael 97–8

Habermas, Jürgen 120
Haine, W. Scott 123
HaLow (802.11ah standard) *see* Wi-Fi HaLow
handiness, and standards 39
Hargreaves, Tom 77
Harper, Richard 78
  The Connected Home 73
Hayes, Vic 20, 39, 41, 42

Healy, Terry, *A History of Intellectual Property in Fifty Objects* 21
Hearst, William Randolph 72
Heer, Tobias 129
Heidegger, Martin 39
hertzian space 24, 113
Hills, Alex 42
  Wi-Fi and the Bad Boys of Radio 19
Hirsch, Eric 55, 56, 61, 68, 69, 79
holistic approach 56
home Wi-Fi *see* household Wi-Fi
homelessness
  and free Wi-Fi 134
Hong Kong protesters (2014) 100
hospitals 143
hotspots, Wi-Fi 115, 118, 134
  growth of 119
  number of 95
  unequal spatial distribution of 133
hotzone 115
household 23–4
  controlling children's ICT use by parents 67–8
  location of technology within 63
  moral economy of the 56, 67, 68
household Wi-Fi 51–81
  and connected home 73–8
  and connectivity 52, 59, 60, 64, 65, 73, 74
  development of standards to increase range and speed 74–5
  domestication approach 53–71, 77, 78–80
    appropriation 56, 61–3
    conversion 56–7, 69–70, 79
    imagination 56, 57–61
    incorporation 56–7, 67–9
    objectification 56, 63–7

growth of 52
number of devices 52, 74
reasons for popularity 52
reconfiguring and transformation of household due to Wi-Fi 65–6, 80–1
and Wi-Fi 6 75–6, 77
Huawei 22
Humphry, Justine 134
Hurricane Sandy (2012) 99

i-Shanghai project 133
iBook *see* Apple iBook
Iceland, home internet use study 70
ICT for development (ICT4D) 97–8
IEEE (Institute of Electrical and Electronics Engineers) 12, 14, 15, 21, 40, 151
standard-setting 19–20, 22, 41, 44, 50
see also 802.11 standards
Île Sans Fil project (Montréal) *see* ISF
imagination (domestication approach) 56, 57–61
Immaterials: Light Painting Wi-Fi project 108, **109**, 110
Inagaki, Nobuya 101
incorporation (domestication approach) 56–7, 67–9
Indigenous governance systems 103–6
infrastructure(s)
common 101
computing 29
definition 26–8
modernity of internet 45–6
provision of comfort and convenience 45
spectrum as 29–33
and Wi-Fi 25–50
innovation
community 95–6
and community Wi-fi networks 94–7
Institute of Electrical and Electronics Engineers *see* IEEE
Interbrand 13
International Monetary Fund (IMF) 10
International Telecommunications Union 31
internet 9–10, 142
access to 10
internet of things 50, 56, 73, 150
iPad *see* Apple iPad
iPhone *see* Apple iPhone
ISF (Île Sans Fil) (Montréal) 88–9, 101, 127, 128

Japan, cafés 123
Jobs, Steve 4, 12, 23, 42
Jones, Matt 90
Jungnickel, Kat 90, 129

K-net 98
Kelty, Chris 93, 94
Kitchin, Rob 138
Kleinman, Sharon 120
Knutsen, Jørn 108
KodaCloud 136
Koolhaas, Jasper 88

Lamarr, Hedy 18, 35
LANs (local area networks) 39, 40, 42, 52, 115
laptops 139
developments in and diffusion of city Wi-Fi 116–19, 139
Laurier, Eric 121
Lee, Christina 25
legal institutions 20
Lemstra, Walter 20, 39, 41
Lenczner, Michael 89
Lennar Corporation 66
Leong, Elaine 80
LinkNYC 136

Linksys 88, 95
Linux 88, 89
Lo Cigno, Renato 101
local area networks *see* LANs
Lofland, Lyn 120–1
London coffee-houses 120–1
London Reform Club 71–2
low-power wide-area network (LPWAN) 136
LTE 146
Lucent Technologies 42

Maccari, Leonardo 101
McConnell, Christopher 133
McCosker, Anthony 122
McDonald's 119
Mackenzie, Adrian 52, 108, 117, 126, 129
McShane, Ian 127, 130–1
magazines, articles on Wi-Fi 57–60
Malaysia, Sarawak 48–9
Marconi radio 30
Marcus, Michael 19, 32, 35–6
Martinussen, Einar Sneve 108
Meinrath, Sascha 99
Mellon, Andrew 42
mesh networks 40–1, 98–100, 115
mesh routers 65
Metcalfe, Bob 36–7
Microsoft 149
Middleton, Catherine 129
MITRE report (1980) 35
mobile internet 5, 118, 141
modems 64
    magazine articles/advertisements on 58
Montréal, Île Sans Fil project 88–9, 101, 127, 128
Morley, David 56, 61, 62, 68, 69, 79
Morse code 30
motor vehicles 143
multi-user multiple-in/multiple-out (MU-MIMO) 75

Mumford, Lewis, *The City in History* 111–12
municipality-driven networks 129
Murray, Alan 133

Nansen, Bjørn 66, 67
Navarro, Leandro 101
NCR Systems Engineering 38, 39, 40, 42, 148
Negroponte, Nicholas 71
neighbour disputes 68
Network Interface Card (NIC) 42
network performance-enhancing products 65
network speeds 17
New York City Wireless 89
New York Times 38, 90
NG Media 98
Nintendo 117
NYC Mesh 99, 100, 103
NYCwireless 87–8, 128

objectification (domestication approach) 56, 63–7
Occupy movement 99
Oldenburg, Ray 120
One Laptop Per Child programme 99
online news services 9
Open Garden 100
open source code 143–4
open source software 82, 92, 93–4
open spectrum 19
open standards 40, 145, 149
Open Technology Initiative 99
orthogonal frequency division multiple access (OFDMA) 75
Oslo, light paintings 108, **109**
Oudshoorn, Nelly 94, 95

Paris Sans Fil 89
PC computers, wireless availability on 42

Index

PC Powerplay 58
Penderecki, Krzysztof 122
Perng, Sung-Yueh 113
personal digital assistants
   (PDAs) 117
Philo, Chris 121
photographs 80–1
platformize Wi-Fi 111, 116, 126,
   138, 139
portable computing/computers
   116–17, 139
positioning software 116
Potts, Jason 129
Powell, Alison 88–9, 127
Powerline adapters 65
privately owned wireless
   networks 128
provider-driven networks 129
pseudorandom radio systems
   34–5
public–private partnerships 128,
   136
public realm, and city life 120
publicly owned wireless
   networks 128
Puregger, Alex 135
Purple WiFi 125, 126

Raby, Fiona 113
radio 19
   pseudorandom systems 34–5
Radio Shack 57, 58
   'Everyday Needs' reference
      guide 58–9
radio spectrum 29–33, 102
   allocation of 6 GHz band
      (2020) 146–7, 148
   authorization of spread
      spectrum by FCC (1985)
      19, 32–3, 45
   regulation of 31–2, 102–3
   see also spread spectrum
random access (direct access)
   36
recipes, family 80
recursive publics 93–4

remote settlements, Wi-Fi in 26,
   46–8
Reply All podcast 25
roads 26–7
round trip time (RTT) 76
routers 52, 53, 63, 76, 78–9, 132,
   143
   advertisements for 60
   appropriation of 62
   bundled with other services
      62, 78
   enhancing connectivity 65
   mesh 65
   placement of within the home
      64
   as residential gateway 79–80
   and round-trip time (RTT) 76
Ruhleder, K. 27–8, 39

Saba, Michael 25
San Francisco Call 30
Sandbye, M. 80
Sandvig, M. 89
Sao Paulo
   WiFi Livre SP 127
Sarawak (Malaysia) 48–9
Sawhney, Harmeet 64
Seattle Wireless 89–90
Sennett, Richard 120
shallow domestication 78
Shanghai 133
Silverstone, Roger 55, 56, 61, 68,
   69, 70, 79
   Television and Everyday Life 57
Skyhook Wireless 116
smart cities 111, 135–8
   critiques of 138
   and Wi-Fi 135–8
   see also city Wi-Fi
smart homes 71–3, 76, 77
   Gates's home 72
   meaning of 72
   precursors to 71–2
smartphones 44, 53, 118, 139,
   145, 152
SNET 86

South Korea 11
Spain
  Catalonian independence
    referendum (2017) 100
spatial frequency re-use 76
Spectator, The 121
spectrum
  as infrastructure 29–33
  regulation of 31–2, 102–3
  release of unlicensed (2020)
    146–7, 148
  spread *see* spread spectrum
  'whitespaces' 74, 102–3
spectrum auctions 32
Spigel, Lynn 69, 71
spread spectrum 34, 41, 42
  authorization of by FCC
    (1985) 19, 32–3, 45
  direct sequence 43
  early experiments in 34–6
  early goal of 35–6
  frequency hopping 43
  and MITRE report (1980) 35
St Cloud (Florida) 128
stadiums 143
standards, Wi-Fi 38–45
  definition 38–9
  developments in to increase
    range and speed of
    domestic Wi-Fi coverage
    74–5
  and handiness 39
  IEEE 802.11 *see* 802.11 *entries*
  open 40, 145, 149
  problem with 41
  setting of by IEEE 19–20, 22,
    41, 44, 50
Star, Susan Leigh 27–8, 39
Starbucks 119
Straubhaar, Joseph 133
Symbol Technologies 40

tablets 118
Taiwanese Wi-Fi use 113
Tallinn (Estonia) 113–14, 124,
  124

Target Wake Time (TWT) 75
Tatler, The 121
teenagers
  and Wi-Fi 70
Telecom Advisory Services 119
telegraph 113
telephone 112
  early history in rural America
    84–5, 87
television 66, 69, 70
  'whitespaces' spectrum 74,
    102–3
Telstra, 5G Wi-Fi Pro 138
Tesla, Nikola 18, 33–4, **34**
'third nature' 113
third sector 84
Titanic, sinking of (1912) 30–1
Torres Strait Islanders 46–7
Townsend, Anthony 87–8
Transmission Control Protocol
  44
Tréguer, Félix 93
TS Wireless 102

UNICEF (United Nations
    International Children's
    Emergency Fund) 3
United States
  history of early telephony in
    rural 84–5, 87
  number of Wi-Fi connected
    devices in 6
  Radio Act (1927) 31
university campuses
  helping to control infection by
    Wi-Fi scenario 151–4
University of Hawaii 18, 36
urban life realms 120
urban Wi-Fi *see* city Wi-Fi
Urry, John 150
user-driven networks 129

Van Oost, Ellen 94, 95
Verhaegh, Stefan 94, 95
Vijn, Marten 88
von Hippel, Eric 95

Waldren, Brian 96
Wang, Mingfeng 133
WANs (wide area networks) 52, 137
warchalking 90–2, **91**
wardriving 90
Wark, McKenzie 112–13
WaveLAN 18, 20, 42, 148
WBA (Wireless Broadband Alliance) 119, 136
WECA (Wireless Ethernet Compatibility Alliance) 13, 42–3 *see also* Wi-Fi Alliance
Werbach, Kevin 31
Wi-Fi
 aspirational dimension of 23, 24, 26, 45, 50
 benefits of 50, 141
 as a brand 12–16
 combination of new and old technologies 16–17
 conception of name 13
 as a connectivity-enabling/gateway technology 51, 62, 69, 73, 74, 135
 contestation of histories 20–2
 distinctive attributes 7
 diversity of histories 18–20
 embedding of in other structures 44, 50
 evolvement of 140–1
 global economic value of 119
 growth in global and domestic population of services 5–6
 hardware and its precursors 33–8
 how it works 52–3
 impact of 5
 as an infrastructure for generating/transmitting data 150
 infrastructural dimensions 25–50
 as a 'last mile' technology 7
 limitations and problems encountered 6, 17, 135–6, 141
 origins 11, 12–13, 18–19
 plasticity of 9–10
 prospects and future 140–54
 reasons for importance 8–9
 reasons for low cost 16
 relationship with 5G 137–8
 as self-provided 28–9, 33, 98
 social futures 151–4
 technologies added to since first version approved 43–4
 as a teleological invention 18
 trademark system used by Wi-Fi Alliance 13–14
 visibility of networks 29
Wi-Fi 4 (802.11n) 44
Wi-Fi 5 (802.11ac) 12, 14, 44, 67
Wi-Fi 6 (802.11ax) 44, 67, 75–6, 77, 131, 136, 142–3, 144, 147–8
Wi-Fi 6E 147–8
Wi-Fi ad-hoc networking 40
Wi-Fi Alliance (was WECA) 5, 13–14, 41, 42–3, 44, 66, 74, 131, 135, 137, 141, 142, 147–8
Wi-Fi Data Elements 143
Wi-Fi devices
 growth in number of 51–2
 number of globally 51, 137
Wi-Fi EasyMesh standard 41
Wi-Fi gradualism 144, 146
Wi-Fi HaLow (802.11ah) 74, 136
Wi-Fi Home Design 66–7
Wi-Fi router *see* routers
wide area networks (WANs) 52, 137
Wifidog 89
Wilken, Rowan 122
Willett, Rebekah 68
Wilson, Charlie 77
WiMax 52
Wired magazine 19
Wireless Broadband Alliance *see* WBA

Wireless Ethernet Compatibility Alliance (later Wi-Fi Alliance) *see* WECA
Wireless Leiden 94–6
Wireless Philadelphia project 129
WirelessLondon 89
World Trade Organization 20
World War II 34–5

Xerox Palo Alto Research Center 36
Xircom 40

YouTube 106

Zigbee 52
Zone Access Public Montréal (was Île Sans Fil) 101